NONLINEAR DYNAMICS IN PHYSIOLOGY

A State-Space Approach

Mark Shelhamer
*The Johns Hopkins University,
School of Medicine, USA*

NONLINEAR DYNAMICS IN PHYSIOLOGY

A State-Space Approach

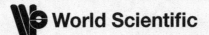

NEW JERSEY · LONDON · SINGAPORE · BEIJING · SHANGHAI · HONG KONG · TAIPEI · CHENNAI

Published by

World Scientific Publishing Co. Pte. Ltd.
5 Toh Tuck Link, Singapore 596224
USA office: 27 Warren Street, Suite 401-402, Hackensack, NJ 07601
UK office: 57 Shelton Street, Covent Garden, London WC2H 9HE

British Library Cataloguing-in-Publication Data
A catalogue record for this book is available from the British Library.

NONLINEAR DYNAMICS IN PHYSIOLOGY
A State-Space Approach

Copyright © 2007 by World Scientific Publishing Co. Pte. Ltd.

All rights reserved. This book, or parts thereof, may not be reproduced in any form or by any means, electronic or mechanical, including photocopying, recording or any information storage and retrieval system now known or to be invented, without written permission from the Publisher.

For photocopying of material in this volume, please pay a copying fee through the Copyright Clearance Center, Inc., 222 Rosewood Drive, Danvers, MA 01923, USA. In this case permission to photocopy is not required from the publisher.

ISBN-13 978-981-270-029-2
ISBN-10 981-270-029-3

Editor: Tjan Kwang Wei

Printed in Singapore

To my family
above all
my wife

Preface

This book is an exploration of mathematics as applied to physiological systems. It is not a mathematics textbook *per se*, nor is it designed to provide a self-contained background in physiology. It is specifically aimed at the interface between these fields, and in particular is intended to address the question: why should I bother with a nonlinear-systems analysis of my physiological system, and how should I go about this analysis? The material, especially in the first 14 chapters, is essentially mathematical. Physiological applications are presented as case studies in later chapters.

It is important to appreciate the role of mathematics in the analysis of physiological systems. In too many cases, mathematical tools (Fourier analysis, nonlinear dynamical analysis, computer modeling) have become so easy to use that they are easily misused[1]. All mathematical tools are based on assumptions, and at the very least the investigator should be aware of these assumptions when using the tools. (The use of statistical t-tests with data drawn from non-normal distributions is a pertinent example from another context.) This book will attempt to provide sufficient mathematical background so that the reader can avoid major mistakes in applying the tools presented here, while keeping the level of rigor firmly in check.

The mathematical approach taken here can be summarized by two well-known quotations. The first is from renowned physicist Richard Feynman: "The next great era of awakening of human intellect may well produce a method of understanding the qualitative content of equations.

---

[1] RM May (2004) Uses and abuses of mathematics in biology. *Science* 303:790-793.

Today we cannot see that the water flow equations contain such things as the barber pole structure of turbulence that one sees between rotating cylinders. Today we cannot see whether Schroedinger's equation contains frogs, musical composers, or morality - or whether it does not."[2]. In fact it is this type of qualitative understanding that is stressed in this book, most obviously through the visualization of system trajectories in state space, and measurements based on these trajectories. The second quote is from mathematician and computer scientist Richard Hamming, in his book on numerical methods: "The purpose of computing is insight, not numbers"[3]. It will be good to bear this in mind, as some of the computational techniques presented here can produce numerical values which might give a false sense of confidence. The field of computational nonlinear dynamics is still at a stage where many of the numerical results should be viewed as estimates, best interpreted as relative values to be compared across conditions or across populations.

The text will also provide some appropriate historical background from time to time. In the vast majority of undergraduate, and even graduate, education, material is presented in the form of revealed truths. Instructors often present the subject matter in a way that makes it seem as if it should be obvious to one and all, if only the student would take the time to think about it for a few minutes. Alas, this is not the way science works. It proceeds in fits and starts, there are debates about what the crucial questions are, there is contention as to the best approaches, and there is often little agreement even when solutions are proposed. This is a human endeavor, and although the end product is extremely reliable, having passed the tests of peer review and trial by fire, the process of getting there is not always straightforward. It is helpful to have a sense of how even today's most "obvious" scientific truths were once muddled in the fog of battle.

This is not solely to indulge the author's personal interest in the history of science. The more important goal is to provide for the reader

---

[2] RP Feynman, RB Leighton, M Sands (1964) The Feynman Lectures on Physics, Vol. II. Reading MA: Addison-Wesley.
[3] RW Hamming (1962) Numerical Methods for Scientists and Engineers. New York: McGraw-Hill.

(especially young students) a sense of the evolution of ideas, the dead ends and unproductive paths, the overall somewhat haphazard nature of the development of a scientific field. This imbues a crucial perspective for anyone considering a career in research. First, it shows (hopefully) the excitement and grand adventure of the research endeavor, and indicates why many of us consider it a true privilege to be a member of the international community that engages in this mad pursuit. Second, it demonstrates that in any scientific field (and especially in one that is new, such nonlinear dynamics in the life sciences), the research literature must be read with a critical eye, until consensus is developed as to the important problems and the legitimate approaches to them. (Of course such consensus is subject to later upheaval, but most of us must for practical purposes be constrained by these limits for a time[4].) Third, the historical perspective should provide some solace to the beginning researcher who feels overwhelmed with the mass of information already available, and must face the feeling that everyone else is faster, smarter, or at least better-funded. Even if these esteemed goals are not met, at least I hope that the occasional historical background will be enjoyable.

Some readers may remember the intense enthusiasm generated in the past with the rise of interest in such areas as neural networks and expert systems in artificial intelligence. There was a great flurry of interest in each of these fields, associated with what we might graciously call "over-enthusiastic claims." In each case, there was great temptation to believe that all existing computational/cognitive/modeling problems could be solved imminently, if only enough support and attention would be devoted to that particular field of endeavor. I recall that in the 1980s the case was made for the creation of distinct Neural Networks Departments in research universities, in order to make clear the separation from electrical engineering and the other progenitor fields. Things have calmed down considerably since that time, and now each field can be seen within a broader perspective. The shortcomings and difficulties of each have become more apparent, and their place in the wider intellectual landscape is both more certain and more circumscribed.

---

[4] TS Kuhn (1996) The Structure of Scientific Revolutions, 3rd edition. Chicago: University of Chicago Press.

And so it is with the topic of this book. In its early years (late 1970s to early 1980s), the promise of the "new" field of nonlinear dynamical systems (more correctly, the re-discovery of computational approaches to nonlinear dynamics) was often vastly overstated. Widely known as "chaos theory" at the time, many in the field were motivated by a mad rush to demonstrate chaotic behavior in a favorite system. With the passage of time, the difficulty of this specific task has become very clear, while at the same time its importance has been called into question. As reflected in the preferred designation of "nonlinear dynamics," the field has attained a broader perspective, improved its techniques, refined the scope of its inquiries, and moderated its rhetoric. In parts of this book this journey will be clear.

The approach taken in this book is based almost solely on analysis of system trajectories in state space. By this I mean that we will consider properties of system behavior that can be derived from these trajectories. This is different from approaches based more directly on time-series analysis or bifurcations of dynamical equations. In particular this is not a mathematics text, nor is it a collection of articles from the research literature. Rather I have tried to tie together the various tools and techniques that have proven most useful to date, and to present the physiological results in a way that emphasizes the system not the tools.

The computational techniques that form the heart of the book are recent by the standards of linear systems theory, but they may appear rather dated in light of the rapid development of nonlinear dynamical analysis. Nevertheless, the methods are by no means outdated. The emphasis on these techniques is deliberate, and based on several factors. First, these techniques are the most firmly established ones, especially for physiological analysis. Second, they serve as good illustrations of the underlying concepts and principles. Third, they are not very difficult to understand and are therefore appropriate for teaching and for developing intuition and insight. Fourth, more recent developments have almost invariably built upon these basic methods. Fifth, they are the methods that I have found to be the most useful in initial dynamical studies of the oculomotor system.

Following on this last point, in many places throughout the text (and most notably in Chapter 5 on dimension estimation), practical suggestions for implementation and interpretation of the computational tools are provided. This information is based on both published recommendations and personal experience. These passages provide the essence of a "user's guide" to some of the tools, which will hopefully make readers more comfortable in attempting their use.

Deciding on the organization of the book was a difficult task. Coverage includes mathematical review material, advanced and recent techniques for analysis of systems and signals, and a review of progress in several different fields of physiology and life sciences. An obvious approach would have been to include pertinent physiological examples in association with the description of each computational technique. The problem with this is that many of the physiological systems have been analyzed with several techniques, and of course each technique has been applied to several different fields of physiology. I have therefore taken a different approach. Part I (Chapters 1-3) introduces basic concepts in signal processing, dynamics, and linear systems. In part II (Chapters 4-14), each chapter is devoted to a single computational technique or dynamical concept; each computational tool is introduced and elucidated with appropriate examples. Then, in part III (Chapters 15-21), different physiological systems are addressed in turn as a set of case studies, and the range of techniques applied to each system can be shown together in a single chapter. This has allowed me to emphasize the physiological interpretations in part III rather than the tools themselves.

The material selected for the case studies consists of examples chosen primarily for their didactic nature. These chapters do not contain comprehensive reviews in a given area of physiology, which would be unwieldy and quickly dated. Rather, they were selected for clarity in application and interpretation, and in some cases novelty and originality. Also note that the methods selected for presentation in the case-study chapters are not necessarily those that have the most promise for future research, or even those that have been most widely embraced, but those that can be understood and appreciated by the intended readers of this book.

Associated with the book's organization is its anticipated appeal to a wide range of readers. The book is suitable as a text for graduate students or advanced undergraduates in such fields as biomedical engineering and neuroscience. However, my hope is that others who wish to acquire a feel for this relatively new approach to physiology will also find the book useful, as a reference for the computational tools, as a primer on the types of physiological questions that can be addressed with these tools, and as an introduction to some key findings in a few physiological systems. These readers might be researchers in a specific field, or those wishing to survey recent advances in nonlinear dynamics. Those specialists interested in a specific physiological system may find the corresponding chapter in section III to be of particular use.

To summarize, my main goal with this book is to provide a volume for reference and review, which can also help to develop intuition on how to approach nonlinear problems in physiology. The book will introduce computational techniques, and show through examples what can be done with them by life-science researchers. My hope is that this combination of reference material and applications will give the reader the background to make use of the material in his or her own research, and the confidence and insight that will be useful for further exploration and invention of new techniques.

I recall vividly the day, when I was in graduate school, when my close friend and fellow student Dan Merfeld came into the lab shortly after reading the book *Chaos: Making a New Science*, by James Gleick. This popular book for non-specialist readers describes the modern rise of computational approaches to nonlinear systems (by and large the material covered in the present book), and gives a taste of some of these approaches. He said, "after reading this, I'm convinced that there are things in our data that we are missing." After reading the book myself, I felt the same way, and my hope is that after reading the book you are holding, you will feel the same way, and in addition will be armed with some of the tools with which to begin your investigation.

In closing, I would like to express my thanks to the many scientific colleagues over the years who have inspired and encouraged me, in particular David Zee, David Robinson, Peter Trillenberg, Larry Young, and Chuck Oman. Dan Kaplan has been most generous in answering my questions on nonlinear dynamics. The students at Johns Hopkins who have taken my course on nonlinear dynamics over the years, and who have asked probing and insightful questions, have helped me greatly in clarifying this material in my own mind. Those deserving of special mention include Scott Molitor, David Scollan, Sarah Plymale, Josh Cysyk, Chris Gross, Nabeel Azar, Olga Telgarska, Wilsaan Joiner, and Faisal Karmali. My dynamical systems research over the years has been supported by several organizations, to whom I am greatly indebted: NASA, The Whitaker Foundation, NSF, NIH (National Eye Institute, National Institute on Deafness and other Communication Disorders, National Institute of Biomedical Imaging and Bioengineering), and the Department of Otolaryngology – Head & Neck Surgery of The Johns Hopkins University School of Medicine.

Thanks are also due to World Scientific for their help in the production of this volume, and Dan Shelhamer for proofreading portions of the manuscript. Portions of section 13.4 are reprinted from the journal *Biological Cybernetics*, volume 93, "Sequences of predictive eye movements form a fractional Brownian series - implications for self-organized criticality in the oculomotor system" (M. Shelhamer), pages 43-53, copyright Springer-Verlag 2005, with kind permission of Springer Science and Business Media. Some material in several other chapters, Chapter 15 in particular, is reprinted from *Journal of Neuroscience Methods*, volume 83, "Nonlinear dynamic systems evaluation of 'rhythmic' eye movements (Optokinetic Nystagmus)" (M. Shelhamer), pages 45-56, copyright 1998, with permission from Elsevier. Portions of chapter 15, including Figures 15.1.1, 15.1.3, and 15.1.6, are reprinted from the journal *Biological Cybernetics*, volume 76, "On the correlation dimension of optokinetic nystagmus eye movements: computational parameters, filtering, nonstationarity, and surrogate data" (M. Shelhamer), pages 237-250, copyright Springer-Verlag 1997, with kind permission of Springer Science and Business Media.

M. Shelhamer
Baltimore MD
March 2006

Contents

Preface vii

1. The mathematical analysis of physiological systems: goals and approaches
   1.1 The goals of mathematical analysis in physiology 2
   1.2 Outline of dynamic systems 5
   1.3 Types of dynamic systems – random, deterministic, linear, nonlinear 8
   1.4 Types of dynamic behaviors – random, fixed point, periodic, quasi-periodic, chaotic 11
   1.5 Follow the "noise" 13
   1.6 Chaos and physiology 14
   General Bibliography 17
   References for Chapter 1 18

2. Fundamental signal processing and analysis concepts and measures
   2.1 Sampled data and continuous distributions 20
   2.2 Basic statistics 21
   2.3 Correlation coefficient 24
   2.4 Linear regression, least-squares, squared-error 25
   2.5 Random processes, white noise, correlated noise 29
   2.6 Autocorrelation 30
   2.7 Concluding remarks 30
   References for Chapter 2 31

3. Analysis approaches based on linear systems
   3.1 Definition and properties of linear systems 32
   3.2 Autocorrelation, cross-correlation, stationarity 33
   3.3 Fourier transforms and spectral analysis 35
   3.4 Examples of autocorrelations and frequency spectra 39
   3.5 Transfer functions of linear systems, Gaussian statistics 43
   References for Chapter 3 44

4. State-space reconstruction
   4.1 State variables, state space 45
   4.2 Time-delay reconstruction 47
   4.3 A digression on topology 50
   4.4 How to do the reconstruction correctly 55
   4.5 Example: detection of fast-phase eye movements 60
   4.6 Historical notes, examples from the literature 63
   4.7 Points for further consideration 66
   References for Chapter 4 70

5. Dimensions
   5.1 Euclidean dimension and topological dimension 74
   5.2 Dimension as a scaling process – coastline length,
         Mandelbrot, fractals, Cantor, Koch 75
   5.3 Box-counting dimension and correlation dimension 81
   5.4 Correlation dimension – how to measure it correctly 85
   5.5 Error bars on dimension estimates 92
   5.6 Interpretation of the dimension 94
   5.7 Tracking dimension over time 96
   5.8 Examples 97
   5.9 Points for further consideration 99
   References for Chapter 5 102

6. Surrogate data
   6.1 The need for surrogates 104
   6.2 Statistical hypothesis testing 105
   6.3 Statistical randomization and its implementation 106
   6.4 Random surrogates 108
   6.5 Phase-randomization surrogate 109
   6.6 AAFT surrogate 110
   6.7 Pseudo-periodic surrogate 113

6.8 First differences and surrogates 114
6.9 Multivariate surrogates 115
6.10 Surrogates tailored to specific physiological hypotheses 117
6.11 Examples of different surrogates 118
6.12 Physiological examples 121
References for Chapter 6 122

7. Nonlinear forecasting
   7.1 Predictability of prototypical systems 124
   7.2 Methodology 126
   7.3 Variations 131
   7.4 Surrogates, global linear forecasting 132
   7.5 Time-reversal and amplitude-reversal for detection of nonlinearity 133
   7.6 Chaos versus colored noise 134
   7.7 Forecasting of neural spike trains and other discrete events 136
   7.8 Examples 137
   References for Chapter 7 139

8. Recurrence analysis
   8.1 Concept and methodology 141
   8.2 Recurrence plots of simple systems 143
   8.3 Recurrence quantification analysis (RQA) 149
   8.4 Extensions 151
   8.5 Examples 152
   References for Chapter 8 153

9. Tests for dynamical interdependence
   9.1 Concepts 156
   9.2 Mutual false nearest neighbors 157
   9.3 Mutual prediction, cross-prediction 160
   9.4 Cross-recurrence, joint recurrence 165
   9.5 Mathematical properties of mappings 169
   9.6 Multivariate surrogates and other test data 170
   9.7 Examples 171
   References for Chapter 9 172

10. Unstable periodic orbits
    10.1 Concepts 175
    10.2 Example 176
    10.3 Physiological examples 178
    References for Chapter 10 179

11. Other approaches based on the state space
    11.1 Properties of mappings 181
    11.2 Parallel flows in state space 183
    11.3 Exceptional events 185
    11.4 Lyapunov exponents 186
    11.5 Deterministic versus stochastic (DVS) analysis 187
    References for Chapter 11 188

12. Poincaré sections, fixed points, and control of chaotic systems
    12.1 Poincaré section 190
    12.2 Fixed points 193
    12.3 Chaos control 203
    12.4 Anticontrol 208
    References for Chapter 12 209

13. Stochastic measures related to nonlinear dynamical concepts
    13.1 Fractal time series, fractional Brownian motion 211
    13.2 fBm, correlation dimension, nonlinear forecasting 214
    13.3 Quantifying fBm: spectrum, autocorrelation, Hurst
          exponent, detrended fluctuation analysis 216
    13.4 Self-organized criticality 217
    References for Chapter 13 218

14. From measurements to models
    14.1 The nature of the problem 220
    14.2 Approaches to nonlinear system identification 221
    14.3 A reasonable compromise 222
    References for Chapter 14 223

15. Case study – oculomotor control
    15.1 Optokinetic nystagmus – dimension, surrogates,
          prediction 225

Recurrence analysis 227
　　　Correlation dimension 229
　　　Surrogate data 230
　　　Filtering 234
　　　Nonlinear forecasting 235
　　　Mutual forecasting 237
　　　Physiological interpretation 239
　　15.2 Eye movements and reading ability 239
　　References for Chapter 15 240

16. Case study – motor control
　　16.1 Postural center of pressure 242
　　16.2 Rhythmic movements 246
　　References for Chapter 16 250

17. Case study – neurological tremor
　　17.1 Physiology background 252
　　17.2 Initial studies – evidence for chaos 253
　　17.3 Later studies – evidence for randomness 256
　　References for Chapter 17 260

18. Case study – neural dynamics and epilepsy
　　18.1 Epilepsy background 262
　　18.2 Initial dynamical studies 263
　　18.3 Dimension as a seizure predictor 265
　　18.4 Dynamical similarity as a seizure predictor 268
　　18.5 Validation with surrogates, comparison of procedures 271
　　References for Chapter 18 273

19. Case study – cardiac dynamics and fibrillation
　　19.1 Heart-rate variability 276
　　19.2 Noisy clock or chaos? 278
　　19.3 Forecasting and chaos 280
　　19.4 Detection of imminent fibrillation: point correlation
　　　　　dimension 283
　　References for Chapter 19 289

20. Case study – epidemiology
    20.1 Background and early approaches 292
    20.2 Nonlinear forecasting of disease epidemics 294
    References for Chapter 20 299

21. Case study – psychology
    21.1 General concepts 302
    21.2 Psychiatric disorders 303
    21.3 Perception and action 306
    References for Chapter 21 309

22. Final remarks
    References on climatic attractors 313
    Suggested references for further study 313

Appendix
    A.1 State-space reconstruction 316
    A.2 Correlation dimension 320
    A.3 Surrogate data 322
    A.4 Forecasting 323
    A.5 Recurrence plots 326
    A.6 Periodic orbits 327
    A.7 Poincaré sections 328
    A.8 Software packages 330
    A.9 Sources of sample data sets 333

Index 335

Chapter 1

The mathematical analysis of physiological systems: goals and approaches

Why should anyone bother with mathematical analysis of a physiological system? This is perhaps as much a philosophical question as a scientific one, and the answer at this point will be somewhat philosophical. We can do no better than to draw on the views of the physicist Eugene Wigner (1960) in his classic essay "The Unreasonable Effectiveness of Mathematics in the Natural Sciences," where he makes the case "that the enormous usefulness of mathematics in the natural sciences is something bordering on the mysterious and that there is no rational explanation for it." Quite a statement from one of the key physicists of the 20th century, but it serves well to show the awesome power of mathematics in describing physical reality.

Of more practical interest, a mathematical description of a system serves to put our knowledge of that system into a rigorous quantitative form, that is subject to rigorous testing (Robinson 1977). In this sense a mathematical model serves as an embodiment of a hypothesis about how a system is constructed or how it functions. The model forces one to focus thinking and make inexact ideas more precise. You may think that you know how a system works, but creating a model of the system truly puts that understanding to the test.

This book will deal with models of systems and their behaviors in a broad sense: is the system random or deterministic, linear or nonlinear, for example. These are broad areas of categorization, but they are also crucial questions that should be addressed before more finely structured hypotheses are evaluated. A major goal in this work is to identify these properties through the use of appropriate computational tools. Answering these questions can not only improve one's understanding of the system but also improve the ability to make diagnoses and apply the appropriate therapeutic intervention in case of pathology.

One type of behavior – *chaotic* behavior – will be of particular interest. The concept of chaotic behavior is appealing because it shows that complex behavior can arise from simple models, lending hope that complex physiological behavior might have simple underlying laws.

1.1 The goals of mathematical analysis in physiology

Anyone reading this book probably does not need to be convinced of the value of a mathematical approach to physiological analysis. Nevertheless an outline of our goals – as related to the tools presented in this book – will give a sense of the direction in which we are heading.

We might list as the first and most fundamental goal that of simply understanding how a system works. This goal is pure and laudable, and corresponds with what many investigators consider to be the defining characteristic of a true scholar: learning for the sake of learning. Of course this is not always appreciated by the public nor by those who fund research, but it is important and undoubtedly continues to be a prime impetus for many researchers, whether admitted or not.

On a more practical (and fundable) level, there is the goal of being able to predict the future behavior of a system. Weather prediction is the prototypical example here, and indeed this example holds a prominent place in the history of nonlinear systems analysis. More correctly, we might say that it has a special place in *modern* nonlinear systems analysis – the use of computers and computational tools to analyze systems that are beyond the reach of conventional analytical methods. By this is meant roughly that they do not have closed-form solutions that can be obtained by solving differential equations, but rather their solutions must be found by approximation methods on a computer. We must be careful to make this distinction because, more than a century ago, mathematicians were already making great strides in the analysis of nonlinear systems. The work of Henri Poincaré is most often cited in this regard. It is often said that Poincaré was the last true mathematical generalist – that is, he was able to grasp and make contributions to all then-extant subfields of mathematics, before specialization became rampant. Some of his most important efforts related to nonlinear systems include work on the fundamentals of topology and celestial mechanics. He was perhaps the first to point out that small uncertainties in the initial conditions can lead to very large uncertainties as a system progresses in time. This is now one of the hallmarks of chaos: sensitive dependence on initial conditions.

He is also, by the way, the author of a set of fine books on the nature of mathematical investigation itself. These books are interesting for the insight they give into the mental processes behind mathematical discoveries. Like watching a great musician or athlete, one might find these works to be either motivational and inspiring, or depressing once the realization sets in that there are people whose abilities are well beyond the upper tail of the normal distribution.

Another distinguished early figure in computational mathematics is John von Neumann, one of the great Hungarian scientists who came to the U.S. early in the 20th century and made major contributions to many fields of mathematical physics, including the calculations behind nuclear weapons and nuclear energy. (This, combined with his ground-breaking work on game theory and his vocal disdain for Soviet communism, made him the model for Dr. Strangelove in the movie of the same name.) von Neumann contributed to the design of the ENIAC, the first electronic digital computer (although the ABC computer by John Atanasoff also has claim to this title), during Word War II. In the course of this work, he thought deeply about the logical foundations of computing itself, and co-wrote a key report that described what has come to be known as the "von Neumann machine": a computer that stores its instructions (software program) internally and makes no hard distinction between stored data and stored instructions, so that the computer itself can operate on its own instructions as easily as it can operate on data (von Neumann 1945). This of course describes virtually every computer that has been made since that time, and has made possible high-level languages and compilers, as just two examples. He was quick to recognize that many intractable problems of mathematical physics could only be solved by recourse to computer approximation methods, and laid out these ideas during the early years of the electronic computer (von Neumann *et al.* 1995, Goldstine 1980).

But let us return to the case of weather prediction. In the early 1960s, Ed Lorenz at MIT was carrying out a numerical investigation (i.e., computer solution) of a set of three partial differential equations that formed a simple model of atmospheric convection. He had run his simulation and found an interesting result, so he re-entered the same starting values and ran the program again. To his surprise, he found that after a short time, the old and new solutions greatly diverged. What had happened? It turns out that when he re-entered the values, he rounded them off to three rather than six decimal places – a seemingly insignificant change, and one certainly below the precision with which meteorological measurements could be made. In any case, it was expected that such a small change in initial conditions might lead to a slowly-increasing difference between the two computer runs, rather than the large non-proportional change that actually occurred. The recognition that this simple nonlinear system could produce such unexpectedly complex behavior is widely credited as the birth of modern chaos theory.

The trajectories of the system's variables were plotted in a state space (this means of examining system behavior is the central theme of this book). This created a finely detailed structure known as an *attractor* for technical reasons that we will cover in due course. The work was published in a classic paper which is often cited but that likely few have read (Lorenz 1963) – indeed it is the much later references to this finding in more general works on nonlinear dynamics to which most writers refer. The results gave rise to the notion of the "butterfly effect," whereby a butterfly flapping its wings in one city today can influence the weather thousands of miles away a week from now. The shape of the Lorenz attractor coincides with the butterfly image. There are many reports of this now-legendary event, including one written by Lorenz himself (Gleick 1988, Lorenz 1996).

From the ability to predict the future course of a system's behavior, the related goal of control follows naturally. Once we can predict the future, it is natural to want to try to control that future, in order to guide the system to a preferred state or keep it away from undesired states.

These basic goals of understanding, prediction, and control, are closely related to the practical clinical goals of diagnosis and treatment, which underlie much of the rationale for research into physiological systems. Understanding a system's behavior and how it is altered under pathological conditions is of course a form of diagnosis, and "control" is effectively just another word for "treatment."

Some of the most advanced work on applying nonlinear system analysis to understanding the course of disease is to be found in the area of epidemiology, while much progress in physiological control is in the area of cardiology. We will return to these topics in later chapters.

Significant progress toward the goals described here can be made with the tools that will be presented in this book, and in particular with the state-space approach that is the guiding theme of the work.

Physiology, and the life sciences in general, has come a long way since the time of an anecdote related by the mathematician Ulam (1976, p. 160). Speaking to a group of biologists, his every attempt to posit some general statement about biology was met with a certain contempt and the reproach that there was an exception in such and such special case. "There was a general distrust or at least a hesitation to formulate anything of even a slightly general nature." Things have changed considerably, although the discovery of overriding principles in the life sciences has not occurred to nearly the extent as in physics, for example. While the computational tools presented in this book can provide some common ground in terms of how to approach some near-universal issues in physiology, the greater hope is that the concepts of nonlinear dynamical analysis themselves will help to establish a common way of *thinking* about such systems (Glass 1991).

1.2 Outline of dynamic systems

By *dynamic system* we mean one that can be defined by a set of variables whose values change with time. (Variables, not parameters – see below.) The variables that thus describe the course of the system's behavior as a function of time are known as *state variables*, because collectively they describe the *state* of the system at any given time. This powerful concept leads us to the *state space*, where the values of the state variables trace out trajectories over time. An example of a dynamic system is the regulation of car speed while driving. Clearly an automatic speed-control system fits the description of a dynamic system, but the interactive "human-in-the-loop" type of manual control is a dynamic system as well. Some of the state variables in this case might be instantaneous speed and foot pressure on the accelerator pedal. These are related in a direct but potentially complicated manner. A simple model might consider speed to be proportional to pedal pressure, while a more realistic model might make this a nonlinear function and include time delays resulting from engine dynamics and neural lag. Even more detailed models could include engine dynamics explicitly, as well as air pressure against the front of the car. Knowing which variables are important to include in the model is one of the keys to successful modeling, and this is in many cases more an art than a science. Indeed the realization that some previously ignored quantity actually is an important state variable is one way to gauge progress in modeling.

We have described some of the state variables of this model. The interactions between the state variables are the essence of the model. Often these interactions are described in terms of differential equations (or difference equations if time is in discrete units rather than continuous). If you know these equations, you have tremendous knowledge about the system. If you do not know these equations, your knowledge is limited, and you typically will have to study and describe the system in terms of the variables that you can measure. This is where the material in this book will be most useful. The ability to proceed from these measurements and analyses on the variables, to the equations, is very difficult in general. We will have occasion to touch on this topic, for which no standard approaches yet exist, in Chapter 14.

The model above describes the mostly mechanical aspect of maintaining a given speed, once you have decided on what speed you want. The model can be expanded to include this decision process, now including such state variables as road conditions, expected time of arrival, confidence in the integrity of the vehicles, and so on. This is a legitimate model, but now the variables become harder to quantify and come under the domain of psychology rather than the physiology of sensorimotor control.

A more purely physiological example is the control of cardiac interbeat intervals (heart rate). Here the important state variables are probably too numerous to list, and include the concentrations of several chemicals in the blood, metabolic rate, mental state, body temperature, and many others. Fortunately, some of these variable are more important than others, especially if the domain of the model is restricted – for example, if it is desired to know the effect of one type of drug or of a specific electrical intervention (external pacing). Such a practical restriction, and the tight coupling between many of the variables, means that significant progress can be made by considering some manageable subset of the entire set of state variables.

With some cleverness, it is possible to describe as dynamic systems many behaviors that at first glance do not appear suited to this approach. Strogatz (1988) gives a playful example of the fateful lovers Romeo and Juliet. One state variable describes Romeo's feelings (love/hate) toward Juliet, and another state variable describes the feelings in the other direction. Based on the strengths of these feelings and the nature of their interactions, various interesting dynamic behaviors can be produced. It is an amusing exercise to consider what other state variables might be involved in this system. (An accessible exposition of this example can also be found in the excellent textbook by the same author (Strogatz 1994).)

As you might imagine, defining the state variables and measuring their values is of crucial importance in modeling and analyzing these systems. In many cases – indeed in most cases that we will consider – it is in fact *not* possible to identify and measure all of these variables. We will study some powerful techniques that will allow us to investigate such systems mathematically even under these restrictive conditions, as long as at least one variable can be measured on the system over time.

In describing and analyzing systems, it is important to distinguish between *variables* and *parameters*. A parameter is a constant – a term in the equations that is fixed, as opposed to the variables, which change as a function of time to reflect the dynamics of the system This issue is closely related to that of statistical stationarity, which loosely speaking means that the statistics of a process do not change with time. If you consider something to be a parameter and in fact it varies over the course of the investigation, then the system has become nonstationary, and you should consider treating the parameter as a state variable.

Probably the best approach to this issue is first to decide on the dynamics that you want to study and model. If a quantity expresses these dynamics – if it changes with time in a manner that reflects these dynamics – then it is a state variable. If that quantity is fixed with respect to the dynamics in question, then it is a parameter. Clearly, what is a parameter and what is a variable depend on what is to be modeled or what behavior is to be analyzed.

As an example, in the case of heart rate, if it is desired to study the effect of a specific short-term physical activity, then other things should be kept constant so that they can be considered as fixed parameters – the subject should not eat during the experiment and it should take place quickly enough so that circadian fluctuations do not have a significant effect. On the other hand, if it is desired to investigate the impact of time of day on the effects of exercise, then time of day becomes a state variable. Time scales such as these constitute one major factor in distinguishing between parameters and state variables.

1.3 Types of dynamic systems – random, deterministic, linear, nonlinear

We can now begin to get to the heart of the matter, by describing and classifying dynamic systems. A very broad categorization, which is nonetheless quite useful for our purposes, considers *randomness* and *linearity*.

While intuitively the notion of randomness is clear to most of us, a rigorous definition is tricky. A *deterministic* system is the opposite of a random system. With a deterministic system, given perfect knowledge of the initial conditions and the system dynamics, the future behavior for all time can be determined. The French mathematician Laplace was a great believer in the concept of determinism as a ruling force in the universe, and made the claim that if given the present positions of all objects and the forces acting on them, their entire future course of motion could be completely and unambiguously determined. He was a true proponent of the power of Newtonian mechanics. Most systems studied by engineers are treated as deterministic, with noise considered as a separate nuisance random process.

The great value of a deterministic system is that, given sufficient knowledge about the dynamics and the values of the state variables at a given time (the state of the system at that time), the future course of the system can be predicted with some degree of accuracy. A random system, on the other hand, is ruled to some extent by chance. Given complete information on the dynamics and initial state, it is not possible to predict precisely the future course of the system (although it may be possible to determine the statistics of the future course – that is, the likelihood of the system being in particular states at particular times). One aspect of the rise of interest in nonlinear systems analysis is the recognition that very complicated behavior – that which is apparently random – can arise from a relatively simple deterministic system. This is another hallmark of chaos, discussed below.

Although noise in the real world is ubiquitous and is the bane of most every experimenter, obtaining true randomness in numerical simulations and computations is much more elusive. This issue arises in the methods covered in this book because it is often necessary to generate random numbers, or to shuffle data into random order, as part of a test for randomness. In the 1950s, the need for such numbers was sufficient that a book was published just for the purpose of providing random numbers (RAND 1955). Of course, computer random number generators (RNG) are available today; these are based on deterministic equations and rely on the fact that they have a very long period – that is, the values they produce will typically repeat, but only after an extremely large quantity of numbers has been generated. In addition, the RNG must be initialized with a "seed" value, and if the seed is not changed the same sequence of "random" numbers will be generated. This can be useful when debugging a program, but for general use some means of randomizing the seed should be used, such as creating it from manipulations on the date and time at the instant that the program is run. A useful source of truly random numbers, which seems to be as random as it gets, is radioactive decay. There are web sites from which one can obtain random numbers based on times of emission of radioactive particles, and such numbers have passed many tests of randomness (see Appendix).

An extension of the points raised above is that no finite system is truly random. Understanding this point is crucial to properly applying the computational methods in this book. A basic concern in many situations is whether a system is random or deterministic. In fact, given a finite data set, a deterministic system can always be constructed that will generate that particular data – the system may have as many parameters or state variables as the data set, but it will reproduce that data set. This of course is a completely uninteresting type of model, and in general we require that a model be able to reproduce a data set while being as simple as possible. This might be a ridiculous example, but the point is subtle, especially in cases where there is a mix of deterministic and random properties.

Let us now move on to linear versus nonlinear systems. A linear system is defined by two properties: scaling and superposition. Scaling means that, if a given input produces a given output, then doubling the size of the input will double the size of the output, and so on for any arbitrary scaling of the input:

$$u(t) \to y(t) \quad \Rightarrow \quad a\,u(t) \to a\,y(t)$$

Superposition means that, if one input produces a given output, and a different input produces another output, providing the sum of these two inputs to the system will produce as output the sum of the two individual outputs:

$$\left. \begin{array}{l} u_1(t) \to y_1(t) \\ u_2(t) \to y_2(t) \end{array} \right\} \Rightarrow [u_1(t) + u_2(t)] \to [y_1(t) + y_2(t)]$$

These properties make linear systems mathematically tractable, and a vast body of work has been devoted to the analysis of linear systems of many kinds. Even so, it should be recognized that, in the real world, there is no such thing as a true linear system. Scaling implies that a system will respond in proportion to the size of the input, but clearly this cannot happen for arbitrarily large inputs. Eventually, an input can be applied that is so large that the system cannot respond in a linear fashion

– the system may saturate or even fail, and in general will produce a distorted output. Thus all physical (and physiological) systems are inherently nonlinear. Nonetheless, there is often a large range of inputs (amplitudes, frequencies, waveforms) for which the system will respond in a linear fashion, or close to it. The trick in carrying out a linear analysis is to know these limits and to restrict the inputs to this range.

In all cases – whether it be an assessment of randomness versus determinism or of linearity versus nonlinearity – one can see that any categorization must take into account the domain over which the analysis is to take place. One must never lose sight of this fact, either during the investigation itself or in its later interpretation.

1.4 Types of dynamic behaviors – random, fixed point, periodic, quasi-periodic, chaotic

Having discussed some ways to categorize *systems*, we now turn our attention to categorizing different system *behaviors*.

Random behavior is, as one might expect, unpredictable. But as mentioned above, once a given data set (typically a time series) has been obtained from a system, the data are no longer random but fixed. Thus the question of randomness in a data series is a relative one, and the methods in this book are designed to give insight into potential mixtures of determinism and randomness. Since noise is present in all physical measurements, determining if randomness is inherent in the system dynamics or in the measurement process is not always straightforward.

Among deterministic behaviors, a *fixed point* is the simplest. It is just what its name implies: a point or state of the system (set of variables) that, once attained, is not departed. A fixed point is sometimes a degenerate state of little interest – a pendulum with a frictional bearing, for example, will have oscillations that slowly decay until the pendulum comes to rest at the fixed point of hanging straight down (zero position, zero velocity), never to move from this state unless externally perturbed. There are more interesting fixed points. One simple nonlinear system is the *logistic equation*:

$$x(n+1) = \mu x(n)[1 - x(n)].$$

This equation is a simplified model of, among other things, the population of an organism from generation to generation, in the face of limited resources. The population at generation n is given by $x(n)$, and the equation shows that the population will grow from one generation to the next until it becomes too large for the available resources (at $x(n)=0.5$), and then it will decrease, until it again is small relative to the resources, and this behavior will alternate (though not always in an orderly way) between these two modes. A fixed point for this system occurs when the population $x(n)$ does not change from one generation to the next: $x(n+1)=x(n)$. The solution to this is easily found: $x^*=1-(1/\mu)$, where the asterisk indicates that it is a fixed point. The dynamics of this extremely simple nonlinear system are fascinating, and some of the original work in bringing this to the attention of the general science community was done by R May (1976).

The next most complicated type of system behavior is *periodicity*. Mathematically, periodicity with period T is expressed as:

$$x(t+T) = x(t).$$

This means that the behavior repeats itself after time T, and then repeats again and again. The classic example of periodic behavior is *simple harmonic motion*, often represented by a theoretical undamped pendulum, which will swing forever if not perturbed.

Nonlinear systems can exhibit a form of periodic motion termed a *limit cycle*, which is an *isolated periodic trajectory* of the state (Strogatz 1994). It can be stable, in which case nearby states are attracted to it, or unstable, so that nearby states are repelled. Various nonlinear oscillators have limit cycles as attractors (Andronov *et al.* 1987). Simple harmonic motion from a linear system, for example from a spring-mass or pendulum, does not produce a limit cycle, since nearby points in state space belong to oscillations of different amplitudes and are neither attracted to nor repelled from the limit cycle: the periodic orbit in this case is not isolated.

A modification of periodic behavior is *quasi-periodicity*, which is a combination of periodic behaviors that have incommensurate periods – that is, their periods cannot be expressed as a ratio of integers. An example is two sine waves, with periods of 1 and $\sqrt{2}$:

$$\sin(t) + \sin(t/\sqrt{2}).$$

This signal is not periodic because, each time the first sine returns to its starting value of zero, the second sine takes on a different value, and the two sines never "match up" because nt and $nt/\sqrt{2}$ (where n is the number of cycles of the first sine) will never be equal to each other for any integer value of n.

Finally, we come to the behavior that motivated this entire field: *chaos*. For a long time – decades if not centuries – the behaviors listed above were recognized as the only ones possible from a dynamic system. Then came Poincaré, and Lorenz and his weather prediction, and nothing has been quite the same since. There are three defining features of chaos:
1. behavior that appears complicated or even random,
2. sensitive dependence on initial conditions,
3. determinism.

The important point is that chaos arises from a deterministic system, yet is so complex in appearance that it might be mistaken for randomness. It is easy to see the importance of this, for once it is recognized that a potentially simple deterministic system can produce complex behavior, a wide array of complex-appearing behaviors come under investigation with a new view toward identifying the underlying system as being deterministic. If the underlying system is deterministic, then it follows rules, and this opens a whole realm of possibilities for understanding and controlling the system.

We are careful in this discussion to distinguish between types of *systems* and types of *behaviors*. There is overlap, of course, as a random system can be expected to produce random behavior. But there is not in all cases a one-to-one correspondence between the two groups, as for example a nonlinear system can produce any of a variety of behaviors such as chaos and fixed points, depending on the system parameters and initial conditions.

1.5 Follow the "noise"

In 1932 Karl Jansky, a communications engineer at Bell Labs (which has since become part of Lucent Technologies) was given the task of finding the source of radio static (noise) in transatlantic telephone circuits (which were carried over shortwave radio at that time, in addition to undersea cable). He determined that, although thunderstorms created much of the noise, there was another source, which varied throughout the day but was not precisely synchronized with the sun. He eventually determined that the source was the Milky Way galaxy, the variation being due to the fact that Earth is not at the center of the galaxy. This finding that stars and other astronomical objects emitted radio waves was the start of the entire field of radio astronomy.

Years later, in 1965, Arno Penzias and Robert Wilson, also at Bell Labs, made a similar discovery. They were looking for the source of noise in some microwave measurements, and they noted a constant low-level signal, without directionality or temporal variation. Little did they know at the time, this turned out to be the "cosmic background radiation" – a remnant of the big bang and convincing evidence for the big bang theory. For this they won a Nobel Prize in 1978.

The point to these stories is this: follow the noise. In these cases, noise turned out indeed to be noise, but the source of the noise turned out to be most interesting and unexpected. In the research that forms the basis of chaos theory, it is often the exploration of variability and apparent noise – random-appearing signals from physical and physiological systems – that has been the starting point for progress.

1.6 Chaos and physiology

Probably the greatest appeal of chaos for physiology is the simple observation that so much physiological activity is highly variable, appearing random or noisy. A chaotic system can appear this way as well, but there is an underlying deterministic structure. The possibility of bringing systematic order to such highly variable phenomena holds great allure and fascination.. The ability to use a set of standard analytical and computational tools to do this only makes the appeal stronger.

Associated with this on the mathematical physics side is the growing realization that even simple nonlinear deterministic systems can exhibit chaotic behavior, and in fact chaos may turn out to be more the rule than the exception in these systems. For example, it has been demonstrated that there are chaotic solutions to cellular membrane equations (Chay 1985, Chay & Rinzel 1985), which has spurred interest in finding experimental demonstrations.

Two main forces drive the interest in chaotic descriptions of human movement control. First, apparently random behavior can arise from non-random systems. Second, variable (random-appearing) behavior can have a functional role (Riley & Turvey 2002). Related to this is the realization – not exactly new but growing – that random noise can have a complex structure of its own, as for example Brownian motion and fractional Brownian motion (Teich & Lowen 2005). There may be optimal levels of random noise and deterministic chaotic dynamics that use the best features of both in tailoring overall system performance.

Conrad (1986) provides a list of some of the possible functional roles for chaos. One is the deliberate generation of diverse behavior, for a number of reasons including the facilitation of exploratory behavior. Another role might be the generation of unpredictable defensive behavior (butterfly motion is given as an example). A third possibility is that it is a metabolically efficient way to generate "noise" or variability, and so prevent the entrainment of different neural structures, so as to maintain flexibility and adaptability.

These ideas all touch on the notion of the "attractor hypothesis" as an appealing paradigm for the rapid and flexible storage and processing of neural information, with advantages such as the avoidance of local minima and systematic exploration. (Similar arguments have been applied to genetic variation and population dynamics: Emlen *et al.* 1998.) The term "attractor" will be defined, with many examples, in Chapter 4. For now, we can think of an attractor as a geometric pattern generated by some measured physiological signal, such as a closed loop generated by plotting velocity versus position during periodic behavior. A more complex example is the higher-dimensional pattern formed by appropriate manipulations of electroencephalogram (EEG) signals. An attractor is an appealing depiction because it represents a type of stable definable behavior that is, if generated by a chaotic system, very flexible and easily altered. Chaotic systems can lead to very interesting attractors.

A paradigmatic though simple study of rhythmic limb movement demonstrates one of the characteristics of studies in this field (Mitra *et al.* 1997). Is the resulting pattern, for example in a graph of velocity versus position, due to an essentially simple periodic oscillation with additive noise, or to an inherently more complicated set of dynamics that are not adequately captured in a simple two-dimensional plot? While the question is still open, and indeed continues to drive much research, it is fruitful to consider the interactions and combinations of variability and determinism (Riley & Turvey 2002), in particular with the view that "more variable does not mean more random." Indeed variation from trial to trial or moment to moment may provide more information than the more time-invariant aspects of a behavior. (Chapter 16 has more on this.)

Another example involves recordings from olfactory bulb EEG (Skarda & Freeman 1987). The authors claim that multiple simultaneous attractors encode different odors, which allows for rapid access to stored odors without the danger of entrainment into local minima. There is rapid convergence of ongoing chaotic activity to an attractor upon inhalation, yet the ability to acquire new odors/attractors. The attractors are not "fixed points" but dynamic entities reflecting continuing neural firing patterns and flexibility even in the case when an odor has been identified. Although the work is based on some simple chaotic measures which might be called into debate in light of subsequent advances, the "chaos paradigm" still resonates with appealing possibilities for the description of brain function. It might be considered a next step in brain modeling, after the digital computer paradigm and the connectionist (neural network) paradigm.

One counterintuitive finding along these lines relates to the "complexity" of cardiovascular dynamics in aging populations (Kaplan *et al.* 1991). At this point we, like the authors of the cited study, use the term "complexity" in a rather loose manner to mean "more noise-like" and "less regular." This study found that complexity decreased with aging. In other words, there is such a thing as "healthy variability." A decrease in this variability can indicate a decrease in health. Based on this and similar results the interpretation came about that variability can endow a system with flexibility and hence the ability to respond and adapt to environmental stressors. (Whether this variability is random or chaotic is a key question which will be addressed in various places in this text.)

Clearly these are largely philosophical issues. They serve to demonstrate some of the great appeal of "chaos theory" in physiological modeling. Our aims in this book are somewhat more mundane. Although we will touch upon larger interpretations when appropriate, and attempt to provide background and historical perspective at times, the main thrust will be on the correct understanding, implementation, and interpretation of computational tools that can help to visualize and quantify nonlinear dynamics in physiological systems. An additional application that has been little explored to date is the use of these computational tools to test and validate mathematical models (Shelhamer & Azar 1997).

One of the hopes of the recent application of nonlinear dynamical methods to physiology is that they can provide a general mathematical framework that has been missing from this traditionally rather qualitative field (Glass 1991). In the main section of this book we will see some tools that can be used to explore noise-like signals, determine if they are truly random or have a deterministic component, and allow us to draw some conclusions about the underlying system. In the rush to apply these methods, there have undoubtedly been many cases in which the apparent ease of the computational procedures has outpaced the users' understanding of the underlying mathematics (May 2004); it is with some hope of preventing further damage along these lines that this book was written.

General Bibliography

HDI Abarbanel (1996) Analysis of Observed Chaotic Data. New York: Springer-Verlag.
PS Addison (1997) Fractals and Chaos: An Illustrated Course. Philadelphia: Institute of Physics Publishing.
DT Kaplan, MI Furman, SM Pincus, SM Ryan, LA Lipsitz, AL Goldberger (1991) Aging and the complexity of cardiovascular dynamics. *Biophysical Journal* 59:945-949.
D Kaplan, L Glass (1995) Understanding Nonlinear Dynamics. New York: Springer.

J Gleick (1988) Chaos: Making a New Science. New York: Penguin.
E Ott, T Sauer, JA Yorke (1994) Coping with Chaos: Analysis of Chaotic Data and The Exploitation of Chaotic Systems. New York: Wiley-Interscience.
M Shelhamer (1998) Nonlinear dynamic systems evaluation of 'rhythmic' eye movements (optokinetic nystagmus). *Journal of Neuroscience Methods* 83:45-56.

References for Chapter 1

AA Andronov, AA Vitt, SE Khaikin (1987) Theory of Oscillators. Mineola NY: Dover.
TR Chay (1985) Chaos in a three-variable model of an excitable cell. *Physica D* 16:233-242.
TR Chay, J Rinzel (1985) Bursting, beating and chaos in an excitable membrane. *Biophysical Journal* 47:357-366.
M Conrad (1986) What is the use of chaos? In: AV Holden (ed) Chaos. Princeton: Princeton University Press, pp 3-14.
JM Emlen, DC Freeman, A Mills, JH Graham (1998) How organisms do the right thing: the attractor hypothesis. *Chaos* 8:717-726.
L Glass (1991) Nonlinear dynamics of physiological function and control. *Chaos* 1:247-250.
HH Goldstine (1980) The Computer from Pascal to von Neumann. Princeton: Princeton University Press.
EN Lorenz (1963) Deterministic non-periodic flow. *Journal of Atmospheric Science* 20:130-141.
EN Lorenz (1996) The Essence of Chaos. Seattle: University of Washington Press.
SB Lowen, MC Teich (2005) Fractal-Based Point Processes. New York: John Wiley & Sons.
R May (1976) Simple mathematical models with very complicated dynamics. *Nature* 261:459-467.
RM May (2004) Uses and abuses of mathematics in biology. *Nature* 303:790-793.
S Mitra, MA Riley, MT Turvey (1997) Chaos in human rhythmic movement. *Journal of Motor Behavior* 29:195-198.
RAND Corporation (1955) A Million Random Digits with 100,000 Normal Deviates. Glencoe, Ill: Free Press.

MA Riley, MT Turvey (2002) Variability and determinism in motor behavior. *Journal of Motor Behavior* 34:99-125.

DA Robinson (1977) Vestibular and optokinetic symbiosis: an example of explaining by modeling. In: R Baker, A Berthoz (eds) Control of Gaze by Brain Stem Neurons, Developments in Neuroscience Vol. 1. Amsterdam: Elsevier/North-Holland Biomedical Press, pp 49-58.

M Shelhamer, N Azar (1997) Using measures of nonlinear dynamics to test a mathematical model of the oculomotor system. In: JM Bower (ed) Computational Neuroscience, Trends in Research. New York: Plenum, pp 833-838.

CA Skarda, WJ Freeman (1987) How brains make chaos in order to make sense of the world. *Behavioral and Brain Sciences* 10:161-195.

S Strogatz (1988) Love affairs and differential equations. *Mathematics Magazine* 61:35.

S Strogatz (1994) Nonlinear Dynamics and Chaos. New York: Addison-Wesley.

SM Ulam (1976) Adventures of a Mathematician. New York: Charles Scribner's Sons.

J von Neumann (1945) First Draft of a Report on the EDVAC. Philadelphia: University of Pennsylvania.

J von Neumann, T Vamos, F Brody (eds) (1995) The Neumann Compendium (World Scientific Series in 20th Century Mathematics, Vol 1). Singapore: World Scientific Publishing Company.

E Wigner (1960) The unreasonable effectiveness of mathematics in the natural sciences. In: Communications in Pure and Applied Mathematics, vol. 13, No. I. New York: John Wiley & Sons.

Chapter 2

Fundamental signal processing and analysis concepts and measures

Before beginning the study of systems analysis *per se*, we first review some basic signal processing concepts and measurements. These topics arise in many areas of analysis and in many guises, but all too often it is simply assumed that they are generally known. This coverage is not meant to be a complete or exhaustive summary, but rather a primer or review of those measures and procedures that will be needed later.

2.1 Sampled data and continuous distributions

In the real world of data analysis (and the focus of this book), one must typically deal with *sampled data*. When discussing the probabilistic or statistical features of these data, it is common to call upon a theoretical construct in which the data are *samples* drawn from a larger underlying *population*. (This population may in reality not exist.)

As an example, suppose that we wish to make some statements about the height of the residents of a certain city. Ideally, we would measure the height of each resident (ignoring uncertainties in these measurements), and then we would have all the necessary data. In practice this is not feasible, and a small sample of the residents is measured. The mean (average) and various other measures can be computed from this set of sample data, and used to make inferences about the population as a whole. But when this is done, it is important to remember that the quantities thus computed are in fact *estimates* of the true values which would result if we had access to the entire population. This is the basis of statistical sampling.

Even if there is no actual underlying population, we can still make use of this model. Suppose, for example, that we measure a time series, such as a set of cardiac inter-beat intervals. We can think of this data set, conceptually, as one particular time series drawn from a larger *ensemble* of possible time series. Obviously, in this case, this underlying population or ensemble does not actually exist, but thinking in terms of this model allows us to use much of the statistical theory that has been developed around it. Measurements such as mean and variance are often made from a single such time series, and used to infer the mean and variance of the total ensemble of processes. This interchange of time and ensemble averaging can be performed, strictly speaking, if the process is *ergodic*, and this is most often assumed to be the case.

2.2 Basic statistics

Let the mean, or average, of a variable $x(t)$ be denoted by $E\{x(t)\}$, where E means "expected value":

$$E\{x\} = \int x p(x) dx.$$

This says that the average value is the sum (integral) of each value multiplied by its probability of occurrence (p). To relate this to the most common application where the number of occurrences rather than their probability is counted, note first that the probability of a specific value of x is the number of occurrences of that specific value divided by the total number of all values and all occurrences of x. Thus we can move from probabilities to counts and vice versa.

The next step is to express the *sample mean* or average in terms of a discrete set of data values, now denoted as $x(i)$ or x_i:

$$\bar{x} = \frac{1}{N} \sum_{i=1}^{N} x(i).$$

A bar over the x indicates mean value, and specifies that we are dealing with a finite sample of data (as would be measured in an actual experiment) rather than a hypothetical continuous random variable $x(t)$ with which much of the theory of random processes is developed.

Although the nomenclature of continuous and discrete (sampled data) variables are interchanged rather loosely in common usage, it is good to keep in mind that, when dealing with actual experimental data, we are dealing with *sample statistics* and not true probability. One implication of this is that, while we can talk about *the expected value* of a random variable as a mathematical certainty with a specific value, when we talk about the sample mean we are dealing with a value that in itself has variability to it – the sample mean is an *estimate* of the true mean or expected value.

One issue that we have neglected so far is the difference between *probability* and *probability density*. In the case of a discrete, or sampled, process, the probability of occurrence of particular values of x has an obvious meaning: it is simply the number of times that value occurs divided by the number of occurrences of all values. For a continuous random variable $x(t)$, the probability of occurrence of a specific value is, strictly speaking, zero. This is because a continuous variable can take on an infinite number of different values, and so the probability of any specific value is some finite quantity divided by infinity, which is zero. Thus, when dealing with a continuous random variable we speak in terms of probability density, and integrate this density between two values in order to arrive at an actual probability. In other words, to find the probability that $x(t)$ is between a and b ($P\{a \leq x \leq b\}$), we would integrate (sum) the probability density between these two values:

$$P\{a \leq x \leq b\} = \int_a^b p(x)dx.$$

We explore these matters in more detail below.

The mean is a measure of *central tendency*; it is, in the absence of other information, what one might consider the "most likely" or "typical" value. The *variance* is a measure of the deviation of the data from the mean value. Again, it can be expressed in terms of the continuous random variable $x(t)$:

$$Var\{x\} = \int [x - E\{x\}]^2 \, p(x)dx.$$

or in terms of the sampled data:

$$\sigma^2 = \frac{1}{N}\sum_{i=1}^{N}[x(i)-\bar{x}]^2.$$

Keep in mind that, as above for the sample mean $\bar{x}$, the sample variance σ^2 is an estimate of the true variance $Var\{x\}$. It turns out that a better estimate of the variance is the slightly modified version:

$$\sigma^2 = \frac{1}{N-1}\sum_{i=1}^{N}[x(i)-\bar{x}]^2.$$

The second version is better because it is statistically *unbiased*. Remember that it is an estimate of the true variance and, based on sampled data, it has a mean and variance itself. The modified version has a mean that is equal to the true variance, and for this reason it is preferable to the first version.

The *standard deviation* is more commonly used than the variance. It is in the same units as the mean, whereas the variance is in those same units squared. The standard deviation is the square root of the variance.

As should be clear from the defining formula, the variance is the average of the deviations from the mean, squared. It is easy to see that this is a measure of deviation from the mean, but why are the deviations squared? Why not define variance simply as the sum of the deviations from the mean? The answer to this is trivial: positive and negative deviations from the mean could cancel each other, leading to misleadingly small values after summing. So why not take the absolute value of the deviations from the mean? There are several reasons why the deviations are squared. First, this gives more emphasis to the larger deviations, which has an intuitive appeal. Second, squaring the deviations makes them positive, so that positive and negative values will not cancel each other. Third, the mathematics of optimization based on minimizing the sum of the squared deviations produces tractable results, as shown below. Finally, the variance as defined here has a natural interpretation with respect to several common probability distributions.

2.3 Correlation coefficient

The correlation coefficient is a measure of the degree of correlation or covariation between two variables. We first provide the "formal" definition in terms of continuous random variables, and then define the more familiar *sample correlation coefficient* that is used with sampled data from real experiments.

We begin by defining the *covariance* of the two random variables x and y:

$$Cov\{x,y\} = E\{[x - E(x)][y - E(y)]\}.$$

This is an extension of the variance defined above, but now the average is found of the deviations of x from its mean, times the deviations of y from its mean. As the name implies, the covariance quantifies the extent to which the two variables x and y vary together. Using some basic facts from probability theory, this can be written as:

$$Cov\{x,y\} = E\{xy\} - E\{x\}E\{y\}.$$

If x and y are *statistically independent,* then the mean of their product is the same as the product of their means – there is no interaction between them:

$$E\{xy\} = E\{x\}E\{y\}.$$

This means that the covariance is zero when x and y are independent. The correlation coefficient is obtained by normalizing the covariance:

$$r\{x,y\} = \frac{Cov\{x,y\}}{\sigma_x \sigma_y}.$$

Let us examine some of the properties of the correlation coefficient r. Think of the two variables x and y as functions of time. Since the covariance is zero when the two variables are independent, so is the correlation coefficient. If y is replaced with a linear function of y, $y'=ay+b$, then the value of r is not changed (a and b are constants). This is easily seen from the defining equation, since the mean of ax is equal to a times the mean of x, and likewise for the standard deviation and covariance: the factor a simply scales the result.

Now, if y is a linear function of x, $y=ax+b$, then the variables x and y are very strongly linearly correlated, and r takes on its maximum value of +1 for $a>0$ and its minimum value of -1 for $a<0$. Therefore r is a measure of the *linear* correlation between two variables. This is a crucial point. As an example, let $x=\cos(t)$ and $y=\sin(t)$, for t over some specified range. No one could deny that x and y are strongly correlated in this case, both being derived from a common variable t by known fixed functions. However, the correlation coefficient is zero. (This is the theoretical value for continuous variables over whole cycles. For actual discrete data the value will be close to zero.) There are many other examples. One could explore this issue very simply by generating a sequence of 100 or so random values, and finding the correlation coefficient between that set of values and the squares of those values. Again these two data sets are intimately related, but the correlation coefficient will not in general be close to 1, and in fact it will vary depending on the actual data values. Nevertheless, r is still a useful tool for quantifying correlations between variables, as long as these properties are kept in mind.

The foregoing discussion refers to the concepts of statistical independence and uncorrelated variables. These are closely related, but they are distinct. Two quantities can be uncorrelated in the sense defined here (a linear measure) and yet be statistically dependent. On the other hand, if two quantities are statistically independent, they will be uncorrelated ($r=0$).

2.4 Linear regression, least-squares, squared-error

Another way the correlation coefficient r arises is in relation to *linear regression*. This procedure was developed in the early 1800s by the mathematicians Legendre and Gauss, who were attempting to determine planetary orbits from noisy (i.e., inexact) astronomical observations.

Let us return to an example from Chapter 1, where the problem of modeling automobile speed control arose. One simplifying assumption in that model was that the relationship between accelerator pedal pressure and car speed was linear: double the pressure and the speed will double, and so on. Assume that, in attempting to test this assumption, you have collected several data points by applying a series of known pressures to the pedal, and measuring the resulting speeds. The data can be graphed as in Fig. 2.4.1.

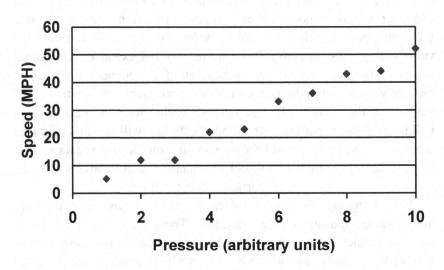

Figure 2.4.1. Car speed as a function of pedal pressure.

We now want to know if, given the unavoidable error that is present in the measurements, this is a linear relationship, and if so, what its parameters are. To do this, we first *assume* that this is a linear relationship, which is eminently reasonable given the nature of the graph. Then we determine a line that describes the data in some "optimal" sense, and finally we determine if this line adequately describes the data in a statistical sense.

Start by assuming a linear relationship between speed (y) and pedal pressure (x): $y=ax+b$. We are faced with the problem of determining values for a and b, given the pairs of data values (x_i, y_i), where $i=1\ldots10$. What exactly do we mean by "optimal" values for a and b? In this case, we will use the same reasoning as in the discussion of the variance of a random variable, and choose to find *estimates* of a and b, denoted by ^, that minimize the sum of the squared errors between the actual data and the linear equation.

$$E = \sum_{i=1}^{N}[\hat{y}_i - y_i]^2 = \sum_{i=1}^{N}[(\hat{a}x_i + \hat{b}) - y_i]^2.$$

Since we wish to find those values of $\hat{a}$ and $\hat{b}$ that produce the smallest error E, we can take the derivative of E with respect to $\hat{a}$ and with respect to $\hat{b}$, set these equal to zero, and solve to find expressions for $\hat{a}$ and $\hat{b}$:

$$\hat{a} = <y> - a<x>$$

$$\hat{b} = \frac{<xy> - <x><y>}{<x^2> - <x>^2}$$

where <·> indicates the sample mean of the values between the brackets:

$$<x> = \frac{1}{N}\sum_{i=1}^{N} x_i \quad <y> = \frac{1}{N}\sum_{i=1}^{N} y_i$$

$$<xy> = \frac{1}{N}\sum_{i=1}^{N} x_i y_i$$

$$<x^2> = \frac{1}{N}\sum_{i=1}^{N} x_i^2 \quad <x>^2 = \left[\frac{1}{N}\sum_{i=1}^{N} x_i\right]^2$$

Note that the formulas for $\hat{a}$ and $\hat{b}$ were derived by minimizing the deviations, or *residuals*, in the vertical direction (i.e., the distances between the actual and fitted y values). The linear regression line does not minimize the distance of each data point from the line (which would be measured perpendicularly to the line), but the vertical distances.

These computations were carried out on the pressure-speed data from above, and the resulting regression line $y = \hat{a}x + \hat{b}$ is plotted in Fig. 2.4.2.

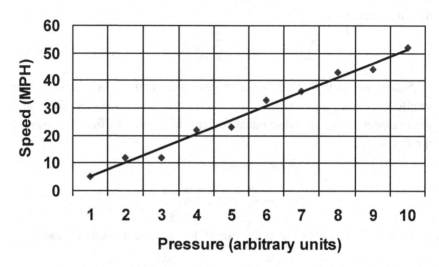

Figure 2.4.2. Data from Fig. 2.4.1 with linear regression line.

How can we measure the quality of this regression line? How well does it fit the data? It is easy to see that the more the data fall along a line, the better the regression quality will be, and so a measure of regression quality will also be a measure of the extent to which the data are linearly related. Not surprisingly, then, the correlation coefficient r can be used to indicate the *goodness of fit* of a linear regression. In this case, $r=0.99$ and the fit is very good.

There is one additional point regarding regression and r with which you should be familiar. Given a value of r computed from the data, how can we know if it is significantly different from zero, in a statistical sense? After all, r is derived from a set of sampled data, which conceptually are drawn from a larger population. Hence r itself is also but a representative value, a specific value based on the specific set of sampled data. Other sets of sampled data would in general produce different values of r. We will dispense with the details here and simply say that it is possible to perform a statistical test on r to see how likely it is that the true correlation differs from zero.

2.5 Random processes, white noise, correlated noise

We mentioned in the opening section of this chapter the notion of an underlying population or ensemble of data series. Then we discussed some aspects of random variables. Here we combine these two concepts by discussing *random processes*. A random process can be thought of as a time series $x(t)$, where each point in time is itself a random variable. That is, the value $x(t_0)$ at time t_0 is a random variable, with a certain probability distribution. This leads us back to our earlier discussion of an ensemble of random processes, where each sample process from the ensemble instantiates a particular value of $x(t_0)$ from this distribution.

A particularly useful random process is *white noise*. This is a random process in which the values at each point in time are statistically independent from each other. The process has no "memory": past values have no impact on subsequent values. One variant of white noise that is very convenient mathematically is *Gaussian white noise* (GWN), which is white noise in which the probability distribution at each point in time is Gaussian (normal). White noise gets its name from the fact that white light is a combination of all other spectral colors of light, and likewise white noise contains all frequencies at equal intensities (obviously a theoretical mathematical construct). We will have more to say about frequency spectra in the next chapter.

In distinction to white noise, there are various forms of *colored noise*. These are random processes in which the values at different points in time are not independent. Any filtering process applied to white noise (or to any other noise) will produce a colored noise. One such digital filter, expressed in discrete time, is:

$$y(i)=[x(i)+x(i-1)+x(i-2)+x(i-3)]/4.$$

This is a *moving average* (MA) filter, so-called because the output value $y(i)$ at time i is the average of the (in this case) four most recent input values $x(\cdot)$. Clearly this will produce a time series y in which the values at consecutive times are related to each other.

2.6 Autocorrelation

As the concept of a random process is an extension of random variables, there is a similar extension to the measurement of correlations. The correlation coefficient r quantifies correlations between random variables, while the *cross-correlation* function $R_{xy}(\tau)$ quantifies the correlation between two random processes $x(t)$ and $y(t)$. Specifically, $R_{xy}(\tau)$ indicates the correlation between $x(t)$ and $y(t)$ when they are shifted relative to each other by a time lag τ.

In our work, it is the *autocorrelation* function rather than the cross-correlation that will be of most use. The autocorrelation is the cross-correlation between a signal and itself. You can think of the autocorrelation function as a string of correlation coefficients, each one indicating how strongly the signal $x(t)$ is correlated with a time-shifted version of itself. Thus white noise (Gaussian or otherwise) has an autocorrelation that has a peak at relative shift, or time lag, $\tau=0$, and immediately drops to zero when τ is not zero: there is perfect correlation of the signal with itself, and there is no correlation at all of the signal with any shifted version. Filtered noise, as well as many other processes and signals, has an autocorrelation that decays more gradually as τ increases from zero. Examples will be seen in the next chapter.

2.7 Concluding remarks

In this chapter we have discussed measures of linearity, and some aspects of random processes. These may seem to be strange topics with which to begin a book on nonlinear deterministic systems. Nevertheless, one must have an understanding of randomness and linearity in order to understand the true meaning of determinism and nonlinearity. (The next chapter covers linear systems in more detail.) Furthermore, the specific aspects presented here are useful in the computational techniques that will be presented for the analysis of nonlinear systems. For example, surrogate data techniques (Chapter 6) entail comparisons with random data, and assessments of the quality of nonlinear forecasting techniques (Chapter 7) make use of linear regressions between predicted and actual values. These types of approaches are of course common features of the quantitative analysis of many systems.

References for Chapter 2

RJ Larsen, ML Marx (2000) An Introduction to Mathematical Statistics and its Applications. New York: Prentice Hall.
A Papoulis, SU Pillai (2001) Probability, Random Variables and Stochastic Processes. New York: McGraw-Hill.
JH Zar (1998) Biostatistical Analysis. New York: Prentice Hall.

Chapter 3

Analysis approaches based on linear systems

Before we can begin to talk in detail about nonlinear systems, we must learn some of the basic approaches to the analysis of linear systems. This will not only make apparent the difficulty of analyzing nonlinear systems, but will also provide us with a useful set of tools to augment the nonlinear techniques. As in Chapter 2, this is not a complete compilation of linear systems techniques, but only a brief review of those matters that will be of most use in applying and interpreting the methods described in the rest of the book.

3.1 Definition and properties of linear systems

What exactly is a linear system? Intuitively most of us will have some sense of a linear system as one that responds in a *proportional* manner to its inputs, and this is in fact the basis for the formal definition. A linear system is defined by two properties:

1. Scaling. If an input $x(t)$ produces an output $y(t)$, then an input scaled by a factor a will produce an output scaled by that same factor: $ax(t) \rightarrow ay(t)$. This means that outputs are proportional to inputs.
2. Superposition: If an input $x_1(t)$ produces an output $y_1(t)$, and a separate input $x_2(t)$ produces an output $y_2(t)$, then the sum of the separate inputs will produce the sum of the separate outputs: $x_1(t)+x_2(t) \rightarrow y_1(t)+y_2(t)$. This means that the two input signals do not interact with each other when going through a linear system. This property allows us to determine the response of a linear system to a very complex input signal by

first determining its response to a number of very simple input signals, which when combined (scaled and added) will produce the more complex input in question. This is the basis for Fourier analysis of linear systems (see below).

A few minutes thought should convince you that there is no such thing as a linear system in the real world. Scaling says that, no matter how large the input signal, we can make it larger and the system will still respond in a proportional manner. Obviously, there will come a point where any real system will not be able to respond to an arbitrarily large input signal, and so all systems will have this type of saturating nonlinearity, among possibly other forms of nonlinearity.

This does not mean that linear analysis is not of great value. It does mean that, when carrying out a linear systems analysis, one must be careful that the system is not being pushed into a range where it can no longer respond in a linear manner. This usually entails restricting the range of input signals in terms of amplitude, and possibly frequency as well. This is part of the *art* of systems analysis.

3.2 Autocorrelation, cross-correlation, stationarity

We introduced the correlation coefficient r in the previous chapter, as a measure of the degree of (linear) correlation between two random variables. This concept can be generalized to measure the degree of correlation between two functions of time, or signals, when one is shifted in time relative to the other.

In analogy to the correlation coefficient, we define the *cross-correlation function* between two functions $x(t)$ and $y(t)$ as:

$$R_{xy}(t_1,t_2) = E[x(t_1)y(t_2)].$$

This defines a function R_{xy} in terms of the functions (signals) x and y at the *specific times* t_1 and t_2. But what does it mean to take the average ($E[\cdot]$) of $x(t_1)$ times $y(t_2)$, at these specific times? Recall from Chapter 2 that x and y are actually *ensembles* of particular processes; think of an underlying population of many different versions of $x(t)$ and $y(t)$. Then, what looks like a single value, $x(t_1)$, is seen as actually many different values, extending over the ensemble rather than over time. Thus the average defined above is an *ensemble average* rather than a time average.

Now, under certain statistical situations (that x and y are jointly stationary, a notion that we will not pursue here), the cross-correlation does not depend on the specific times t_1 and t_2, but only on their difference:

$$R_{xy}(\tau) = R_{xy}(t_2 - t_1) = R_{xy}(t_1, t_2).$$

The *autocorrelation function* will be more widely used in our work than the cross-correlation. The autocorrelation is the cross-correlation of a signal $x(t)$ with itself (i.e., the cross-correlation R_{xy} with $y(t)=x(t)$):

$$R_{xx}(\tau) = E[x(t)x(t+\tau)].$$

This is sometimes normalized to have a value of 1.0 at $\tau=0$, in analogy with the correlation coefficient r which has a value of 1.0 for perfectly correlated variables. Examples of autocorrelation functions are given below in section 3.4.

A rigorous definition of autocorrelation as defined above must include the notion of a *wide sense stationary* (WSS) process. WSS has two defining features. First, the mean is independent of time:

$$E[x(t)] = \mu \text{ (a constant), for all } t.$$

Second, the autocorrelation function depends only on the difference between the two time values, as we implicitly stated above:

$$R_{xx}(t_1, t_2) = R_{xx}(t_1 - t_2) = R_{xx}(\tau).$$

This definition of WSS effectively means that statistical moments of the process $x(t)$ up to second order (mean, variance, autocorrelation, *etc.*) do not depend on time t. This assumption is very commonly made in statistical signal processing applications, and we make use of it in most of our applications as well. (One case in which the assumption is violated is that of fractional Brownian motion, which we will see in Chapter 13.)

3.3 Fourier transforms and spectral analysis

A consequence of linearity is one extremely nice property: sinusoidal signals do not have their frequencies modified by a linear system. If the input is $x(t)=\sin(\omega t)$, then the output will be of the form $y(t)=A\sin(\omega t+\phi)$. The input sine has been modified by the system in two ways: its amplitude has been changed by a factor A and its phase has been changed by an amount ϕ. The frequency ω is unaffected.

It is this property, and the property of superposition noted above, that underlies the power of the Fourier analysis of signals through linear systems. If we can describe how a linear system alters the amplitude and phase of an input sinusoid, and if we can decompose a more complicated input signal into a sum of sinusoids, then we know everything we need in order to determine the response of the system to that complicated input. Conceptually, we pass each individual sinusoid through the system and add the resulting outputs to obtain the output that would occur if the complicated input had been presented. The depiction of a linear system's sinusoidal response in terms of amplitude and phase is known as a *transfer function*. In graphical form it is a *Bode plot*. The fact that a great many complicated signals can be decomposed into sums (or integrals) of sinusoids was demonstrated by Fourier in the 19th century. We will not pursue this aspect of linear systems further, as there are many books available that describe such analysis. We take pains to point out, however, that although we have pointed out the benefits of Fourier analysis for linear systems, we can in fact find the frequency content of almost any arbitrary signal via Fourier analysis. Whether the signal is applied to a linear system or not is of no concern to the Fourier transform.

The *Fourier transform* is defined mathematically as:

$$X(\omega) = \int_{-\infty}^{\infty} x(t)e^{-i\omega t}dt = \int_{-\infty}^{\infty} x(t)\exp(-i\omega t)dt,$$

where ω is the frequency in radians/sec (ω=2πf, where f is frequency in Hz) and i is the square root of -1. This transform is often derived from the *Fourier series* representation, where the expression of a periodic signal x(t) (with period T, ω₀=1/T) as a sum of sinusoids is more apparent:

$$x(t) = \sum_{k=0}^{\infty} a_k \cos(k\omega_0 t) + \sum_{k=0}^{\infty} b_k \sin(k\omega_0 t).$$

Here it is clear that x(t) is the summation of sines and cosines of different frequencies, and in general an infinite number of terms is required to reproduce x(t) perfectly. The coefficients of the expansion are given by:

$$a_0 = \frac{1}{T}\int_0^T x(t)dt, \; a_k = \frac{2}{T}\int_0^T x(t)\cos(k\omega_0 t)dt, \; b_k = \frac{2}{T}\int_0^T x(t)\sin(k\omega_0 t)dt.$$

The term a_0 is the DC (zero-frequency) level of x(t). (Readers familiar with the fact that $e^{i\omega t} = \cos(\omega t) + i\sin(\omega t)$ can begin to see how the integral definition can be derived from the series representation.)

This form is useful when dealing with mathematical manipulations and continuous-time signals, but when dealing with sampled data it is the *discrete Fourier transform* (DFT) that is more useful:

$$X(k\Delta\omega) = \sum_{j=0}^{N-1} x(j) e^{-i2\pi k\Delta\omega j\Delta t} = \sum_{j=0}^{N-1} x(j)\exp(-i2\pi k\Delta\omega j\Delta t).$$

Here, the spectrum X is defined only at discrete frequencies kΔω, which are multiples of a fundamental frequency Δω. In other words, both the time series x and the spectrum X are in discrete form. The DFT is most often computed with a very efficient algorithm known as the *Fast Fourier Transform* (FFT), which takes into account trigonometric symmetries in the DFT to reduce the number of individual computations.

One problem with the FFT is that it is so easy to use. It is easy to fool yourself into thinking that you are doing a legitimate spectral analysis by simply applying the FFT to your data. There are pitfalls, however, which we can only touch upon here. One which most users appear to be familiar with is the choice of appropriate sampling frequency when digitizing the time series; this should be more than twice the highest frequency in the signal, and is easily accomplished by analog filtering of the signal before digitizing. A more subtle problem is that the DFT (and by extension the FFT) makes the implicit assumption that the time series is periodic, with a period as long as the time series itself. (This is implicit in the fact that sine and cosine functions are used to approximate the input time series, and these functions are periodic.) In other words, given a series $x(i)$ of 1000 points, the DFT treats this as an infinitely long signal that repeats itself every 1000 points. If the amplitudes at the beginning and end of the signal are very different ($x(N){\neq}x(1)$), then a discontinuity exists in this implicit periodic signal. Since a discontinuity has a large amplitude change in a very short time, an artifactual high-frequency transient is introduced, which will contaminate the computed spectrum. One of the ways around this is to "window" the data: multiply the signal $x(i)$ by a function that tapers to zero at the beginning and end, to eliminate this type of discontinuity. Windowing has other valuable properties, and much has been written on the overall subject of "window carpentry," the design of windows with desirable features. Another way to approach the problem of transients in the time domain, or signals that begin and end at very different amplitudes, is to use a discrete approximation to the continuous Fourier transform, but an approximation that is not the DFT and so does not make the same periodicity assumption as the DFT (Harris 1998).

Comparison of the defining equations for the Fourier transform and the cross-correlation is revealing. The Fourier transform at a particular frequency is essentially a cross-correlation between the time series $x(t)$ or $x(i)$ and a sinusoid (sines and cosines or a complex exponential) at that same frequency. In this sense we see that the Fourier transform indicates how well the input signal resembles a set of sines and cosines at different frequencies.

The reason that the transform contains terms derived from sines and cosines, rather than only sines, is to be able to represent the phase offsets of the different sinusoidal components as well as their amplitudes. A simple trigonometric identity shows that a sine wave with arbitrary phase offset can be expressed as the sum of a sine wave and a cosine wave at the same frequency, with an appropriate ratio between the amplitudes. This is how the Fourier series above, expressed in terms of sines and cosines, can represent sine waves with different phases. Another way to make this representation is with complex exponentials and Euler's equation:

$$e^{i\omega t} = \exp(i\omega t) = \cos(\omega t) + i\sin(\omega t), \quad i = \sqrt{-1}.$$

The details of this representation can be found in many books on linear systems or signal processing.

We have alluded to the fact that the frequency components that make up a Fourier spectrum have different amplitudes and phases. It is this ability to decompose a signal into sine waves of appropriate amplitudes and phases that is the essence of Fourier analysis. However, the equations given above provide Fourier coefficients in a form that is not so directly useful. We can find the *magnitude spectrum*, which gives the amplitude of each of the component frequencies. For the Fourier series, it is:

$$|X(k)| = \sqrt{(a_k^2 + b_k^2)}$$

at each frequency k. For the DFT it is:

$$|X(k\Delta\omega)| = \sqrt{\text{Re}[X(k\Delta\omega)]^2 + \text{Im}[X(k\Delta\omega)]^2}$$

where Re[·] and Im[·] denote the real and imaginary parts of the complex quantity, respectively.

Similarly we can find the phase spectrum:

$$\angle X(k) = \tan^{-1}[b_k / a_k]$$

or:

$$\angle X(k\Delta\omega) = \tan^{-1}[\text{Im}[X(k\Delta\omega)] / \text{Re}[X(k\Delta\omega)]].$$

The magnitude spectrum is very often expressed in terms of power rather than amplitudes. This is the square of the amplitude values given previously. A fact that we will use later is the relationship between autocorrelation and power spectrum: the power spectrum $S_{xx}(\omega)$ or $S_{xx}(f)$ is the Fourier transform of the autocorrelation function. Note that the power spectrum does not contain any information on the phases of the frequency components, only on their amplitudes. Likewise, the autocorrelation function does not contain phase information. Thus, many different time signals $x(t)$ can have the same autocorrelation and power spectrum; they will differ in their phase spectra. We make use of this fact in the generation of *surrogate data* in Chapter 6.

3.4 Examples of autocorrelations and frequency spectra

Autocorrelation functions and power spectra for several different time signals are shown in Fig. 3.4.1. In the left column are time series, in the center column are autocorrelations, and in the right column are corresponding power spectra. These represent the broad range of spectral types normally seen in physiological systems.

The autocorrelation function of a 5 Hz sine wave is a cosine function at the same frequency. This is periodic, since a sine wave is perfectly correlated with a shifted version of itself every time the shift is equal to one period of the sinusoid. The corresponding spectrum is very nearly a single spike at the frequency of the sine. In general there may be some *leakage* of this single frequency to adjacent frequencies in the DFT (FFT). This may occur if an integer number of sinusoidal cycles is not contained in the time-domain signal, and can be reduced by windowing the data as discussed above. Another factor contributing to potential leakage is the fact that the sinusoidal frequency of the time signal might not coincide exactly with one of the discrete frequencies computed by the DFT; the power of the sinusoid is then "split" between the adjacent DFT component frequencies. These effects are common in spectral analysis of this sort. They would be easy to see in the case of a simple sine wave, but they are easily overlooked in more complex cases.

The next example is the spectrum of a 5 Hz square wave, which is mathematically the sum of sines at odd harmonics of the fundamental frequency. This is very clear from the spectrum, where small spikes are present at 15, 25, and 35 Hz; since the amplitudes of the harmonics decay rapidly (as $1/k$ where k is the number of the harmonic) they are not easily seen beyond that. Again, the autocorrelation reflects the fact that the signal is perfectly correlated with itself after every shift of one period.

The third example is a physiologic waveform: optokinetic nystagmus (OKN). This is a signal that we will see several times throughout this book. It is a signal that represents horizontal eye position, in response to a wide-field random visual scene that moves uniformly at constant speed in one direction. The eyes follow the moving pattern in one direction (slow phases) and intermittently move back rapidly (fast phases) to pick up and follow a new part of the pattern. The autocorrelation indicates that the signal has a strong nearly periodic nature, due to the fact that fast phases occur at about three per second. The spectrum has a dominant peak at about 3 Hz, corresponding to this near-periodic component. Since the signal is not strictly periodic, however, there is significant spectral energy at other frequencies as well. (The time signal has a large offset, such that the average value is not zero. This leads to a large spectral component at 0 Hz, which has been removed from the graph to make it easier to see the details of the rest of the spectrum.)

In these three examples, the spectra have been plotted on linear axes. In cases of more complex spectra, it is common to use logarithmic axes for both magnitude (spectral power) and frequency, as in the next three examples.

The next example, at the top of Fig. 3.4.1b, is Gaussian white noise (GWN). The time signal shows a great deal of variability, as expected. The autocorrelation is very nearly zero (completely uncorrelated) except at zero lag when the correlation is perfect. The power spectrum, in theory, is completely flat – constant power at all frequencies. That is approximately the case here, but since this is one sample from an ensemble of possible processes, we observe some statistical deviation from the ideal. If a large number of white noise processes were generated and their spectra averaged, the result would be much more like the theoretical ideal.

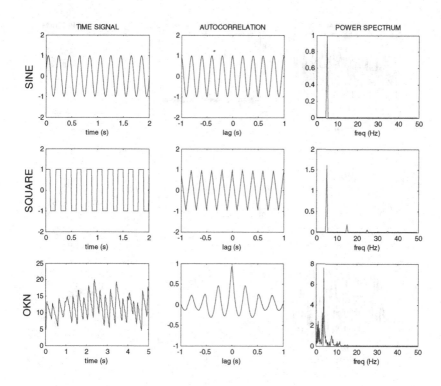

Figure 3.4.1a. Examples of signals in the time domain (left column), and corresponding autocorrelation functions (center column) and power spectra (right column). From the top, the signals are a sine wave, a square wave, and a type of reflexive eye movement known as optokinetic nystagmus (OKN).

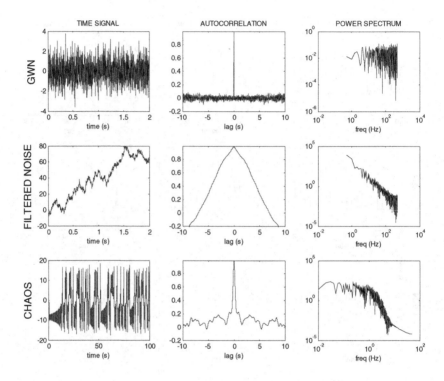

Figure 3.4.1b. Examples of signals in the time domain (left column), and corresponding autocorrelation functions (center column) and power spectra (right column). From the top, the signals are Gaussian white noise (GWN), filtered noise, and a variable from the chaotic Lorenz system.

The example below GWN is a type of filtered noise. It is GWN that has been passed through an integrator: each value is the sum of all the previous values from a GWN signal. This is a strong form of low-pass filtering; as seen in the time series, high-frequency variations have been substantially reduced. This is apparent as well in the spectrum, which drops in magnitude as frequency increases. (The specific shape of the spectrum has a decay rate that is proportional to the frequency squared: $S_{xx}(f) \sim 1/f^2$. This is because an integrator is represented by $1/s$ in Laplace notation, and since s represents frequency and power spectrum is proportional to amplitude squared, the decay rate is proportional to $1/s^2$ or $1/f^2$.) The autocorrelation shows that the signal is well-correlated with itself at small time shifts, and these correlations decay very slowly.

The final example is from a chaotic system, the Lorenz system described previously. The time series is the progression of one of the three state variables of the system, which is known to be chaotic. The autocorrelation decays relatively rapidly, showing that the signal has no strong periodicity, but it does have some regularity because some of the features of the time signal are repetitive. The spectrum has a complex shape, but eventually decays rapidly with frequency. It was thought for a time that this type of $1/f^\alpha$ shape was a sure sign of chaotic behavior, but many examples have been found of systems that generate signals of this general shape but which are not chaotic. We will not use spectral analysis alone as an indicator of chaos.

3.5 Transfer functions of linear systems, Gaussian statistics

One more point regarding spectral representations is of interest. If the transfer function of a system (the description of how the system modifies the amplitudes and phases of sinusoidal inputs) is $H(f)$, and the power spectrum of an input signal is $S_{xx}(f)$, then the power spectrum of the resulting output signal is:

$$S_{yy}(f) = S_{xx}(f) \times |H(f)|^2.$$

Although we have not covered Gaussian statistics in any detail, there is one further property that will be useful to us. The sum of any number of Gaussian random variables is itself a Gaussian random variable. The mean and variance will in general change due to summation, but the probability distribution will remain Gaussian. This has a practical implication when dealing with linear systems: if you apply a Gaussian signal to a linear system, you will get a Gaussian signal out.

The especially nice amalgamation of the properties of linear systems, Gaussian statistics, and least-squares regression has led to one incarnation of the *Engineer's Prayer*: "Lord, please make the world linear, quadratic, and Gaussian."

Alas we will find that it is not to be.

References for Chapter 3

M Akay (1994) Biomedical Signal Processing. New York: Academic Press.
EN Bruce (2001) Biomedical Signal Processing and Signal Modeling. New York: John Wiley & Sons.
CM Harris (1998) The Fourier analysis of biological transients. *Journal of Neuroscience Methods* 83:15-34.
AV Oppenheim, RW Schafer, JR Buck (1999) Discrete-Time Signal Processing (2nd Edition). New York: Prentice Hall.
SW Smith (1999) The Scientist and Engineer's Guide to Digital Signal Processing. San Diego: California Technical Publishing. [Available free on-line at http://www.dspguide.com/]

Chapter 4

State-space reconstruction

Having established some of the basics of linear systems analysis, we are now ready to move on to the analysis of nonlinear systems. We begin by describing, in some detail, the graphical-computational technique that underlies almost all of the subsequent methodology that we will study. It is based on the concept of state variables introduced in Chapter 1. The concepts and techniques presented in this chapter form a crucial basis for the subsequent material in this book.

4.1 State variables, state space

State variables were introduced in Chapter 1 as those quantities that change over time and reflect the behavior of a dynamic system. The concept of a *state space* was also mentioned briefly. This is a mathematical construction, in which each state variable is plotted along one of the axes.

For a simple example of state space, consider a damped oscillator, such as a pendulum with friction. We represent the angular position of the pendulum at time t as $x(t)$. The amplitude of the sinusoidal oscillation decays exponentially:

$$x(t) = e^{-at}\sin(\omega t).$$

To create the state space for this system, we must first identify some appropriate state variables. It turns out that position and velocity are ideal variables in this case, with velocity given by:

$$\dot{x}(t) = \omega e^{-at}\cos(\omega t) - ae^{-at}\sin(\omega t).$$

We will simplify the math for this example by setting $a=1$ and $\omega=20$. Since ω is much larger than a we can approximate and ignore the second term in the derivative above:

$$\dot{x}(t) \approx \omega e^{-at} \cos(\omega t).$$

Now we can create the trajectory in the state space, by plotting $x(t)$ and its derivative, as in the right side of Fig. 4.1.1. The time series $x(t)$ is plotted to the left.

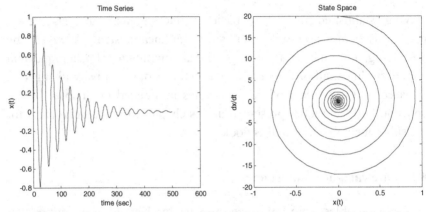

Figure 4.1.1. Damped oscillation. Time series (left) and state space trajectory (right).

It is easy to understand the dynamics of the system from this state space trajectory. The trajectory begins at position $x(t)=0$ and velocity $\dot{x}(t)=20$, its maximum value. As the sinusoid decays in amplitude, the trajectory spirals in toward (0,0), when the oscillations stop.

This simple example reveals several general aspects of representation in the state space, sometimes also called the *phase plane*. No matter what the initial state, this system will always come to rest at (0,0); this point is called an *attractor*, since it attracts all trajectories. In general an attractor can be much more complex than this simple point, and many research papers speak loosely of "state-space attractors" when referring to sets of trajectories, whether or not they have been demonstrated to be true attractors. One experimental approach to confirming that a physical system contains an attractor is to perturb the system and see if the trajectories return to some subset of the state space (Roux et al. 1983). Another feature of this simple example is that the trajectory does not cross itself. This in fact is a general feature of the state space for a deterministic system. If the trajectory crosses itself, and if the state of the system at a given moment is at the point of intersection, then it cannot be determined which path the trajectory will follow. This contradicts the concept of determinism. Therefore, if trajectories appear to cross, the system is either random, or the dimension of the state space is not high enough to depict the trajectories accurately. The second issue is one that will occupy much of our attention, in this and subsequent chapters.

4.2 Time-delay reconstruction

The example given above neatly avoided some crucial issues. How do we know what to use as state variables? How many do we need? What if we cannot measure the state variables from an actual physical or physiological system? Fortunately we have a way to address these issues. The technique is *time-delay reconstruction* or *time-delay embedding*, and its development is one of the keys to the resurgence of interest in computational approaches to nonlinear dynamics in the 1980s.

Time-delay reconstruction is almost absurdly simple yet extremely powerful. Instead of using the actual state variables, such as $x(t)$ and its derivatives, we use successively delayed values of $x(t)$. Examples are given in Fig. 4.1.2, based on the damped oscillator. In each graph, the values of $x(t)$ are plotted versus delayed values of $x(t)$. In other words, the points on each graph are given by $(x(t), x(t-\Delta t))$. The left graph is for a time delay of 5 and the right graph for a delay of 20.

One important point should be immediately obvious: the trajectories resemble the one obtained above that used the actual state variables. A more precise meaning for "resembles" is outside of the scope of this text, but we will discuss it briefly in the next section. Suffice it to say at this time that all of the plotted trajectories spiral in to a fixed point at the origin, and they do so without any self-crossings. The exact shapes of the plots are slightly different, depending on the value chosen for the time-delay parameter in the reconstruction.

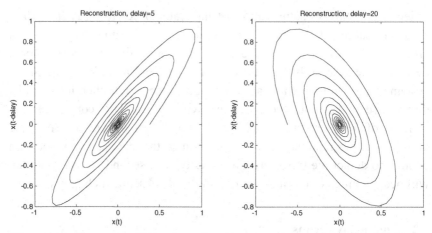

Figure 4.1.2. Damped oscillation from Fig. 4.1.1, shown in the state space using time-delay reconstruction with a delay of 5 (left) and a delay of 20 (right).

If the original data points in the time series are represented by $x(i)$, then the reconstructed attractor consists of the M-dimensional points $y(i)$, generated from the time series as follows:

$$y(1) = [\, x(1),\ x(1+L),\ ...,\ x(1+(M-1)L)\,]$$
$$y(2) = [\, x(1+J),\ x(1+J+L),\ ...,\ x(1+J+(M-1)L)\,]$$
$$y(N) = [\, x(1+(N-1)J),\ x(1+(N-1)J+L),\ ...,\ x(1+(N-1)J+(M-1)L)\,].$$

Here N is the number of points on the reconstructed trajectory or attractor. M is the embedding dimension (the number of "state variables"); each trajectory point $y(i)$ is composed of M values from the time series $x(i)$, separated by time L. (The M-dimensional points are sometimes referred to as "vectors" since they can be treated as vectors in an M-dimensional space.) J is the interval between first elements of successive attractor points, and is usually set to 1.

A simple example will make these equations more clear. Assume a time signal $x(i) = \{1,2,3,4,5,6,7,8,9\}$, and let $M=3$, $L=2$, and $J=1$. Then the points on the trajectory are:

$$y(1) = [x(1)\ x(4)\ x(7)] = [1\ 4\ 7]$$
$$y(2) = [x(2)\ x(5)\ x(8)] = [2\ 5\ 8]$$
$$y(3) = [x(3)\ x(6)\ x(9)] = [3\ 6\ 9].$$

Although we refer to this as "time-delay" reconstruction, the formulation presented here actually represents "time-ahead" reconstruction. That is, the successive elements of each point $y(i)$ are further ahead in time rather than behind in time. This makes no difference in the application of the methods in this book, but can lead to some confusion since it will still be referred to as "time-delay" reconstruction but the formulation presented here will be used.

Proper selection of the parameters M and especially L turns out to be of prime importance in using this technique correctly; this is covered below. One might ask why this technique works – how does it preserve the essential dynamics of the system? We can make a plausibility argument. Recall that the definition of a derivative is:

$$f' = \frac{df(x)}{dt} = \lim_{h \to 0} \frac{f(x+h) - f(x)}{h}.$$

This definition shows clearly that the derivative is based on the difference between two values of the function $f(x)$ spaced close together in time. The time-delay reconstruction can be considered a form of this differencing operation, for a fixed and finite (rather than small and decreasing) value of the time difference (h in this case). (One might think from this definition that very small values of the time delay L are preferred in the reconstruction, but this is not typically the case, as we will see.) We have noted previously that in identifying state variables for a system, it is common to use successive derivatives of some measured quantity: position, velocity, acceleration, and so on. In many cases it can be demonstrated that these derivatives do indeed form a legitimate set of state variables, in the sense that they give complete information about the system state at all times. Since the successive derivatives are good state variables, and the use of consecutively time-delayed values approximates the differencing operation that defines a derivative, we can see that the use of time-delay reconstruction is a plausible means to recreate a set of trajectories in state space.

An extremely attractive feature of the use of time-delayed values is that we can work in high-dimensional state spaces ($M=10$ is not untypical) without having to identify a large number of state variables and without having to take higher-order derivatives. The latter point is especially important, since differentiation is a process that almost always enhances noise in real data. This is because the transfer function of a derivative with respect to time is $H(s)=s$, and so the magnitude of the transfer function increases without bound as frequency increases (the power spectrum is $S_{xx}(f)=f^2$); since noise is typically a high-frequency phenomenon, differentiation amplifies the noise, oftentimes disproportionately to the signal of interest. Thus time-delay reconstruction allows us to deal with many state variables without introducing noise.

One other seemingly magical aspect of the reconstruction should be addressed. The method allows us to obtain complete information on the state of the system through the measurement of a single variable. This can only be accomplished if all of the true state variables of the system are coupled in some way. There cannot be an isolated subset of states that evolves with a separate set of dynamics, independent of the dynamics that affect the variable that is measured and used in the reconstruction.

The mathematical relationship between the classical concept of *observability* (the ability to reconstruct the system state from output measurements) and time-delay reconstruction has been investigated (Letellier *et al.* 2005).

4.3 A digression on topology

We have given a plausibility argument, but by no means a proof, for why time-delay reconstruction works. Such a proof is well beyond the scope of this book, and involves mathematical details of differential geometry, manifolds, and topology. Nevertheless a short digression on the pertinent topology will give some flavor of the mathematical proof.

The basis of the proof is the Whitney embedding theorem (Whitney 1936). In order to begin even our vastly simplified understanding of this theorem, we start with a few definitions from differential topology, which encompasses the study of calculus on multi-dimensional curved surfaces and spaces (as opposed to three-dimensional "flat" Euclidean space).

A *manifold* is a space with specific mathematical characteristics. It is sometimes easiest to think of a manifold as a curved surface (such as a saddle) within space, but in fact defining certain properties and operations on the surface make it into a *space* in the technical sense. This space (surface) can be of any finite dimension. A space is a manifold if, near every point in that space, the local region resembles Euclidean space (i.e., a local coordinate system can be defined). This means that we can specify locations and directions on the manifold. The concepts that follow from this allow us (or rather mathematicians) to perform calculus on manifolds, by making rigorous such concepts as smoothness, continuity, and inverse.

Now assume that we have two such spaces or manifolds, X and Y, and a function f that *maps* points in X to points in Y. This function is called a *homeomorphism* if it has these properties: 1) each point in X is associated with one and only one point in Y (f is single-valued and has a single-valued inverse f^{-1}), 2) f is continuous, and 3) f^{-1} is continuous. By *continuous* we mean, intuitively, that the function is smooth (there are no abrupt breaks or jumps in the function).

To help solidify the concept of homeomorphism, some simple examples are illustrated in Fig. 4.3.1. Each graph shows a time-delay reconstruction of the damped oscillation from Fig. 4.1.1, after it has gone through a mapping function f. The reconstruction is carried out in each case with a delay of 20 as on the right side of Fig. 4.1.2. Four different mapping functions are examined. Although these examples are described in terms of a simple function of two variables (the function acts on the x and y values of points on the original trajectory), a more inclusive way to think of this operation is that the function maps the first space (where the reconstructed trajectory exists) to a new space, by stretching or otherwise modifying the original space in one or more directions. The nature of these modifications determines whether or not the function or map is a homeomorphism.

In panel A is the reconstruction after the map or transformation $(x_1,x_2) \rightarrow (x_1,5x_2)$ has been applied to the data; in other words, the values along the ordinate have been scaled up by a factor of 5. This simply stretches the trajectories vertically, as seen by comparison with

Fig. 4.1.2. The two-valued function f in this case is a homeomorphism, since all three properties given above apply.

Panel B shows the reconstruction for the function $(x_1,x_2) \to (x_1,|x_2|)$. This function is *not* a homeomorphism since, by virtue of the absolute value, two different points in the original trajectory space can be mapped to a single point in the new space. This leads to trajectory crossings in the new space. The function shown in panel C is $(x_1,x_2) \to (x_1,x_2^2)$, which is also not a homeomorphism, for the same reason. In panel D is the function $(x_1,x_2) \to (x_1,x_2^3)$, which is a homeomorphism; although the remapped trajectories are compressed near the origin, there are no crossings, and since raising a number to the third power is continuous and invertible, the mapping is homeomorphic.

The point of these examples is to develop some intuition as to what a homeomorphism looks like in this two-dimensional space. Although a homeomorphic map or function might distort the object (the trajectory) or the space in some way, it retains the overall shape or essential nature of the original. Abrupt alterations, tearing, rending, or folding are not allowed. There is a sense in which graphs A and D in Fig. 4.3.1 are more like Fig. 4.1.2 than are graphs B and C; the former are homeomorphisms.

A *diffeomorphism* is a homeomorphism that preserves differentials (or increments, or derivatives): the mapping function f is differentiable and so is its inverse. In other words, there is no loss of differential structure in the mapping function. Loosely speaking, the *flow* of a trajectory – its motion in different directions at different points – is preserved in such a mapping.

An *embedding* is a function or mapping from the space X to the space Y that is a diffeomorphism and is also smooth (derivatives exist at all points). An embedding is also of course a homeomorphism, which means that the topological properties (such as connectedness and neighborhoods of points) of the two spaces are identical.

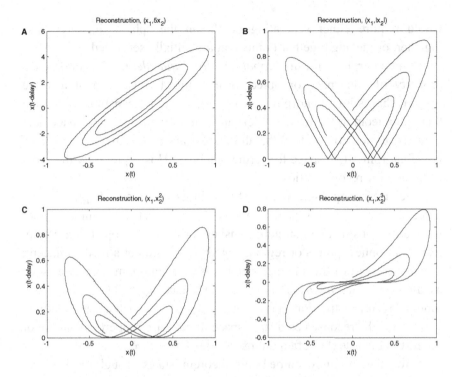

Figure 4.3.1. Examples of the effects of four different mapping functions. In each case, the indicated function is applied to the damped oscillation of Figure 4.1.2, and a time-delay reconstruction of the transformed data is graphed. The transformations or maps in panels A and D are homeomorphisms.

As an aside, we note that the spaces we are dealing with (topological spaces) surprisingly do not require a measurement specification – distances *per se* between points are not important. Rather, the crucial element is the notion of the *neighborhood* of a point – is that point connected to other points or are there separate and disjoint subsets of points? Thus it is *connectivity* and not *distance* that is a primary topological characteristic. As an example, there is a homeomorphism between any closed curve and a circle. The surface of a cube is homeomorphic to that of a sphere. In each case, a "smooth" distortion can change one into the other. A homeomorphism will not, on the other hand, make the surface of a sphere into a donut (torus), because to do so would require either separating points on the sphere that are close together, or placing together points that are initially separated.

Homeomorphic means *topologically equivalent*. A topological property of one space or object is unchanged (invariant) in a space (object) that is homeomorphic to the first. So, if we can find something that is homeomorphic to the underlying unobserved true state space, and if it is easier to deal with, then we can make use of the new homeomorphic state space for further analysis. This is precisely the point of time-delay reconstruction.

One useful topological property is dimension (see Chapter 5). As noted above, a topological property is one that is retained under transformations that can stretch, translate, or rotate an object, but cannot tear it or connect points or regions together that are not already adjacent. The critical part of this characterization is that important properties such as dimension are *invariant* under a topological embedding, and since time-delay reconstruction (performed properly) is such an embedding, we can use the reconstructed state space to measure properties that are of interest in the true (but often unobservable) state space.

Now, the Whitney embedding theorem states (roughly) that any smooth m-dimensional manifold can be embedded in a conventional Euclidean space of dimension $2m$. In other words, any m-dimensional surface, no matter how complex, can be transformed into an object in familiar Euclidean space, and this transformed object will be topologically equivalent to the original. The new space (the embedding space) must be at least twice the dimension of the original space in order to guarantee a proper embedding (i.e., to guarantee that the transformation is a homeomorphism).

Finally, we come to the specific theorem that makes time-delay reconstruction mathematically valid: the *Takens embedding theorem* (Takens 1981). After the above, this theorem will appear anti-climactic, although it is of prime importance. It says, basically, that almost any function that meets certain very broad criteria can serve as an embedding function or map f. These criteria are met by the time-delay reconstruction technique. Therefore, time-delay reconstruction creates a proper embedding from the original trajectory (actually, from the manifold on which the attractor trajectories lay) to the conventional Euclidean space that we can work with. For this reason the method is also referred to as *time-delay embedding*. The reconstructed trajectory (or attractor) is topologically equivalent to the original if the embedding space has a dimension of at least $2m+1$, where m is the dimension of the original space in which the attractor trajectories exist.

Such is the magic of time-delay reconstruction. Rest assured that it is not in any way necessary to understand the details of the proof outlined here in order to make use of the method, any more than it is necessary to understand complex analysis in order to make use of Fourier transforms and the frequency domain. The beauty of the method is that it allows us to visualize and work in spaces of high dimension using simple mathematics (functions applied to multi-dimensional vectors).

Proof of the validity of time-delay reconstruction assumes access to an infinite amount of noise-free data. Needless to say, we do not run into such a case very often in the laboratory. Thus, the practical application of time-delay reconstruction is much less straightforward than one might hope. Faced with finite, noisy data from real-world experiments, how can we know that we have a proper embedding? This question has occupied a vast amount to time and effort, and we take it up in the next section.

4.4 How to do the reconstruction correctly

Carrying out a proper time-delay embedding or reconstruction, with real data, can be thorny. The key parameters in the process are the dimension of the embedding space (the *embedding dimension M*), and the time delay L. In other words, how many state variables should there be, and how far apart in time should be the delayed elements of each point in the state space. We address each question in turn.

The key idea in the choice of time delay L is that the elements that make up an attractor point $y(i)$ should be close enough in time that they loosely approximate a derivative and are dynamically related, yet far enough apart in time that they are not repetitive. Each point $y(i)$ should capture some dynamic information about the system, and if the elements $x(i)$ of that point are too close together, the information they provide will be redundant.

One of the simplest and yet most effective suggestions for choosing L is that it should be a small multiple (2 or 3) of the *correlation time* of the signal $x(t)$. The correlation time is the time shift τ at which the autocorrelation function $R_{xx}(\tau)$ of the time series $x(t)$ has decayed to $1/e$ of its peak value. This is one way to quantify the notion that the consecutive $x(i)$ values should be far enough apart in time to be somewhat but not completely independent (uncorrelated). This simple rule of thumb is a good starting point for the selection of L.

Since the autocorrelation assesses linear correlations (see Chapter 2), a more general form of this same concept has been suggested, based on *mutual information* (Fraser & Swinney 1986). Mutual information is a statistical measure from information theory that indicates the amount of "information" about one random variable that is given by knowledge of another random variable. (By *information* we refer here to a specific technical definition, which is a measure of the *entropy* or statistical variation in a set of values.) Mutual information is a way to quantify the question: what does the distribution of $x(i)$ tell us about the distribution of $x(i+L)$? The mutual information between a time series $x(i)$ and its shifted version $x(i+L)$ is computed for various values of L until the mutual information is minimized. This value of L will provide consecutively delayed values of $x(i)$ for the reconstruction that are independent and therefore, presumably, will provide the most overall information in the coordinates of the points $y(i)$ on the attractor.

Although promising in theory, mutual information is not trivial to calculate and its utility has been called into question (Martinerie *et al.* 1992). Judicious application of the autocorrelation criterion, with two- and three-dimensional plotting of the attractors reconstructed with candidate values of time delay L, seems to suffice in a great many cases.

Selection of embedding dimension M is the other major issue. The embedding dimension M should be large enough that the attractor is properly embedded in the topological sense. In particular, there should be no trajectory crossings if the system is truly deterministic (although noise of various types can introduce apparent intersections which can often be safely ignored).

A promising approach to both of these questions, which enjoys widespread use, is that of *false nearest neighbors* (FNN: Kennel *et al.*

1992). If an attractor is reconstructed in an embedding space with too small a dimension M, then points on the attractor that are actually far apart in space will appear artificially close together – the trajectories are compressed because the embedding space is not big enough for them to fully expand. These points that appear close together in M dimensions but are actually far apart in a higher-dimensional space are *false neighbors*. FNN quantifies this concept.

To make this explicit, let us first define the distance between two points $y(i)$ and $y(j)$ in M-dimensional space:

$$D_M(i,j) = \sqrt{\sum_{k=1}^{M}[y_k(i) - y_k(j)]^2} \; .$$

Here, D_M denotes the distance as measured in M dimensions, that is, with M delayed elements in each point $y(i)$ and $y(j)$. The subscript k to the right indicates that corresponding delayed elements are subtracted from each other in the distance calculation. This is nothing more than the well-known Euclidean distance measure, extended to M dimensions.

A point $y(j)$ is a false nearest neighbor of $y(i)$ if the distance between the two in $M+1$ dimensions is much greater than the distance in M dimensions:

$$\frac{D_{M+1}(i,j)}{D_M(i,j)} > R_{thr}$$

Here, R_{thr} is a distance-ratio threshold. If the distance increases by more than this factor, then we call the corresponding points *false neighbors*. A value of approximately 10 for R_{thr} is suitable in many cases.

In operation, an initial value for embedding dimension M is set. Then, each point on the attractor is taken in turn as a reference point. The nearest neighbor to each reference point is found by computing the distance in M-dimensional space between the reference and every other point and identifying the minimum distance. Then, the distance between these same two points in found in $M+1$ dimensions. If the ratio of these two distances is greater than R_{thr}, the points are false nearest neighbors. Across all reference points, the proportion of nearest neighbors that are false nearest neighbors is found, for the given dimension M. Then M is increased by one, and the process repeated. Finally, the proportion of false nearest neighbors can be plotted as a function of M.

Another criterion must be added for cases, such as noise, where the "nearest neighbors" are not in fact "near" to each other. Given a reference point, its nearest neighbor may be only "accidentally" near, rather than because of inherent dynamics and the geometry of a deterministic attractor. For any arbitrary set of points in M-dimensional space, a simple calculation reveals that, averaged across all pairs of points, the mean increase in inter-point distance that results from increasing the dimension to $M+1$ is approximately two times the variance of the element being added to the distance computation, which is $y_{M+1}(i)$. But $y_{M+1}(i)$ is just the original time series $x(i)$, delayed by an amount ML. Therefore, an additional criterion for considering two points to be false nearest neighbors is if they are nearest neighbors in M dimensions, and the distance between them in $M+1$ dimensions is much greater than the mean inter-point distance for an arbitrary distribution of points:

$$\frac{D_{M+1}(i,j)}{D_A} > R_{A,thr}, D_A = \text{variance of } x(i).$$

Another way to address this same problem (the nearest neighbor not being really near) might be to specify a maximum distance criterion: points cannot be nearest neighbors unless they are closer together than this criterion distance, which could be specified as some small fraction of the largest inter-point distance (a measure of the maximum extent of the attractor). This has apparently not been investigated.

The FNN concept is elucidated in Fig. 4.4.1, using only the first criterion in the discussion above. A time-delay reconstruction of the Lorenz attractor, using the x variable, is given at top left ($M=2$). Two points that are very close to each other in this two-dimensional reconstruction are marked by filled circles, at (0,2) on the graph (the points are so close together that the two circles overlap and appear as one). In the graph to the right, the attractor is reconstructed in a three-dimensional embedding space ($M=3$), and the two points are now clearly distinguishable. A slight tilt of the attractor in the third dimension has revealed that these points are not, in three dimensions, as close together as they appear to be in two dimensions. Since in two dimensions there are no points closer to either of these than they are to each other, they are *nearest neighbors*, and since they are artificially close together and spring apart in a higher-dimensional embedding space, they are in actuality *false nearest neighbors*. The graph at the bottom shows the proportion of nearest neighbors that are false nearest neighbors, as embedding dimension increases. Beyond the degenerate case of $M=1$ the proportion decreases rapidly, and by $M=3$ there are few if any false nearest neighbors.

This result clearly indicates the value of the FNN method, but it also reveals one other practical aspect of attractor reconstruction. The embedding theorems guarantee that an embedding dimension of $2d+1$ is sufficient to reconstruct an attractor of dimension d, but in practice this can be accomplished with a much smaller embedding dimension. In this case, it is known that the Lorenz attractor has a true dimension of slightly greater than 2 (we will discuss non-integer dimensions in the next chapter), and so $M=5$ is guaranteed to produce a proper embedding, but in fact the number of FNN indicates that $M=3$ may be sufficient in this case to "disentangle" the attractor.

All of these various approaches to setting values for the parameters of the time-delay reconstruction are useful, and in particular the false nearest neighbor algorithm is an excellent example of the way in which simple mathematical manipulations can give insight into situations in high-dimensional space. Even with these methods, however, a "brute force" approach is often the most convincing: whatever is being measured (e.g., attractor dimension) should not change as embedding dimension is increased, or with small changes in time delay. Robustness of the quantity that is derived from the reconstructed attractor is still one of the most compelling arguments that the reconstruction was performed properly.

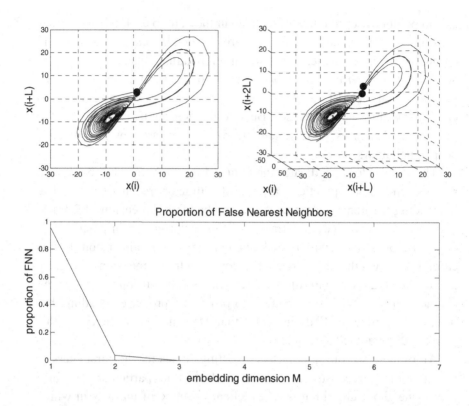

Figure 4.4.1. False nearest neighbor method applied to the Lorenz attractor. Top left: attractor reconstructed with $M=2$. Two points that are nearest neighbors are indicated with (overlapped) large dots. Top right: same attractor reconstructed with $M=3$. Two nearest neighbors are now distinguishable. Bottom: proportion of false nearest neighbors as a function of the embedding dimension M.

It turns out that, at least for dimension computations, embedding dimension and time delay are related, as we will discover in the next chapter.

4.5 Example: detection of fast-phase eye movements

While not directly related to the dynamics of a system in the sense that has been discussed in the previous chapters, the example presented here shows how time-delay reconstruction can be used in a signal-processing task. The example we use is the identification of fast phases in a type of eye movement known as *optokinetic nystagmus* (OKN). As show in Fig. 4.5.1A, OKN is a rhythmic (but not periodic) eye movement that is induced by presenting a wide-field visual scene that moves uniformly in one direction at a constant velocity. The eyes follow the visual scene on one direction with *slow phases*. Intermittently, the eyes move rapidly in the other direction with *fast phases,* to pick up and follow a new point on the scene. The alternation of slow phases and fast phases creates the characteristic sawtooth waveform called *nystagmus*. Note that there is considerable variability in OKN: the fast phases do not begin or end at the same locations, the slow phases are not all of the same duration, the fast phases are not all of the same amplitude. This variability is an issue addressed later in Chapter 15.

Here, we wish to perform a simple signal-processing task: automated identification of the fast phases. In clinical and research settings, various OKN parameters are of interest which require separation of the slow and fast phases; examples are slow-phase velocities, fast-phase intervals, and fast-phase durations. These parameters can only be measured after the slow and fast phases have been parsed. In this specific example, the identification problem is not challenging, and a simple velocity-threshold algorithm can perform the task very well. Nevertheless, to show a simple application of time-delay reconstruction, and to provide a framework for fast-phase identification in more difficult nystagmus signals, the problem was addressed via time-delay reconstruction of the state space (Shelhamer & Zalewski 2001).

A time-delay reconstruction of the OKN signal is shown in Fig. 4.5.1B and C, viewed from two different angles. Since the signal was sampled uniformly at 500 Hz, the data points are 2 msec apart and their spacing can be used to judge eye velocity at various points on the trajectory. During slow phases, the points are close together and often form a continuous line. During the fast phases, the points are farther apart. In this state space, it appears that the slow phases are approximately aligned along a sheet or plane, with the fast phases projecting out of and back into that plane. We hypothesize that in even higher dimensions, the grouping of slow phases in a (hyper) plane might be even stronger.

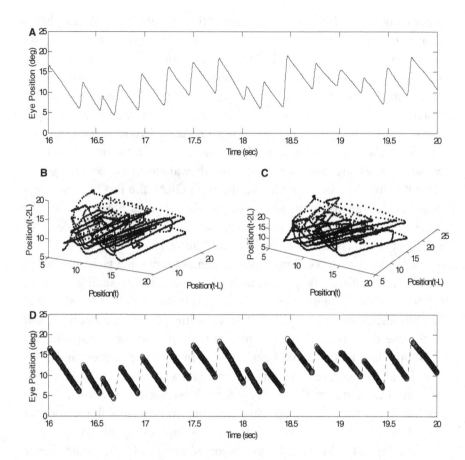

Figure 4.5.1. A) OKN eye-movement signal. B,C) Reconstruction of OKN trajectories with M=3, viewed from two different angles. Slow phases, where consecutive points come close together in time, are roughly aligned, suggesting that a high-dimensional plane might contain them and not the fast phases. D) Performance of algorithm to identify fast and slow phases. Slow phases so identified are plotted with circles.

This suggests an algorithm for identifying the fast phases. First, manually select a few slow phases based on observation of the OKN signal. Then, use time-delay reconstruction to create the OKN trajectory in an M-dimensional state space. Fit a plane of dimension M-1 to those slow phases identified manually. Consider as fast phases any points that deviate from the fitted plane by more than a given amount. Based on this identification of fast phases, fit a new plane to the remaining slow-phase points. Repeat the process, each time fitting a plane and rejecting points off of that plane, until no improvement is seen. Those points in and near the fitted plane belong to slow phases, and points farther from the plane belong to fast phases. An example of the performance of the algorithm is shown in Fig. 4.5.1D.

It is obvious that several details have been left out of the description of this algorithm, such as the parameters of the time-delay reconstruction (M, L, *etc.*) and the distance threshold. However, it is not the intent here to describe the algorithm in detail, but rather to show one way in which time-delay reconstruction can be used to gain intuition about a problem and generate a solution.

4.6 Historical notes, examples from the literature

Analysis of systems in the state-space has a long history, both mathematically and physically. We can trace the fundamental concepts back to the work of Descartes, who (legend has it) while watching a fly buzz about, had the insight to realize that the location of the fly at a given point in time could be specified by three numbers, each giving the location along one of three directions in three-dimensional space. Hence was born the Cartesian coordinate system. We now would say that the fly's position can be given by a three-dimensional state vector. (A more general state vector would also include the fly's *attitude*: in which direction it was pointed, as specified by angles with respect to the three coordinate axes. The motion of the fly, if deterministic, could then be given by a *state-transition matrix*, which operates on the state vector to produce future values of the state vector, and hence the future course of motion. This is the basis of much of linear system analysis in the state space, and it is one reason that *linear algebra* is called what it is.) The beauty of the Cartesian conceptualization is that well-developed and powerful algebraic rules for the manipulation of numerical quantities could now be used to describe and analyze geometrical objects and motions in space: analytical geometry.

Many other contributions were made along these line through the years, but a special place in this history is reserved for Poincaré, who, as noted in Chapter 1, contributed greatly to the mathematics of dynamic systems. Among other things, he envisioned the motion of trajectories in state space in analogy to fluid flow. He also developed the idea of the *Poincaré section*, in which a plane is positioned to intersect the state-space trajectories, and the points of intersection on this plane are analyzed for dynamical properties. This effectively reduces the dimensionality of the analysis by one.

In a more applied vein, in the 1880s the French scientist Lissajous used light reflected from a mirror attached to vibrating objects to study the vibrations. Years later, his name is used to denote the patterns on an oscilloscope screen when two different signals are applied to the horizontal and vertical inputs; the shape of these *Lissajous figures* can be used to determine such things as the relative phase of the two signals. This is a two-dimensional state-space representation.

Coming to more modern approaches in the area of nonlinear dynamics and "chaos," a key paper is that of Packard and colleagues (1980). This was apparently the first publication to suggest that time-delayed values could be used to reconstruct a state space. Although this is a widely cited paper that can be credited with introducing the time-delay approach, both this paper and David Ruelle (1990) point out that it was actually Ruelle's suggestion to use consecutively time-delayed values to reproduce the state space. (Before their work in dynamical systems, some of the authors of this paper attempted to use early microprocessors worn on the body to predict the outcome of a wheel spin in roulette, based on initial measurements of ball and wheel speeds and positions. They have since gone on to apply methods of nonlinear forecasting – a topic that we cover in Chapter 7 – to financial trading. These adventures are reported in two books by Bass (1985, 1999).)

Using this time-delay method, a later paper (Roux *et al.* 1983) produced and analyzed the state space trajectories for the *Belousov-Zhabotinski reaction*, which consists of more than 30 chemical components and maintains non-equilibrium spatial and temporal oscillations for long periods of time. Time-delay reconstruction was performed on measurements of the concentration of one of the chemicals in the reaction. In a non-periodic state, the frequency spectrum of the time series of one of the reactants indicates broadband noise, suggesting that the system is random. But the state-space trajectories form patterns that instead suggest that the system, even in its non-equilibrium non-periodic state, is deterministic. A *Poincaré section* was created by "slicing" the attractor with a two-dimensional plane and plotting on that plane the points where consecutive trajectories intersect it. A smooth function could be fit to these intersection points, again suggesting that the underlying dynamics are deterministic. (A shortcoming of this study is that there is no indication as to how the plane that forms the Poincaré section was chosen, and how representative of the entire attractor this particular Poincaré section might be.) From the Poincaré section, a *Lyapunov exponent* was computed, which indicates whether consecutive trajectories are pulled closer together or are thrown farther apart, the latter indicative of a chaotic system. This computation was "alarmingly sensitive" to the data and to some of the computational parameters. It is because of this and similar findings in some other studies that we will not make use of Lyapunov exponents in this book. There are more recent studies that suggest that reasonable Lyapunov exponent estimates can be produced, if great care is exercised.

In order to be termed an "attractor," the state-space trajectories should converge to a well-defined region or object in the space. This was verified in the Roux study by applying mechanical and chemical perturbations to the reaction, and observing the behavior of the resulting trajectories. Since they indeed converged to a well-defined region, the term "attractor" could be correctly used to describe the grouping of the trajectories. (The reference to "strange attractor" in the title of this article refers to the fact that the attractor from a chaotic system is a *fractal*: it has infinitely detailed fine structure. Such a fractal attractor is termed a *strange attractor*. See Chapter 5 for more on this topic.)

The critical issues of time delay and embedding dimension were dealt with in this study as well, but not in detail. Selection of the delay time L was addressed simply by plotting the state-space trajectories with different values, and seeing for which values the most structure in the attractor was revealed – a completely subjective approach that is nonetheless still valid. Selection of embedding dimension M was addressed in a similar manner. In principle, the attractor can be reconstructed in increasingly higher dimensional spaces "until additional structure fails to appear." In the next chapter we will see one way to do this objectively, but in this study this was only approached subjectively, by noting that "the character of the attractor is clear" with $M=2$. Nonetheless, trajectory intersections are easily seen in the two-dimensional plots in the paper, and so an embedding dimension of at least $M=3$ is required. We note that this is considerably less than the dimension of $2N+1=61$ that is guaranteed to be sufficient by the Takens theorem ($N=30$ being the number of independent chemical components, and therefore theoretical state variables, in the reaction).

4.7 Points for further consideration

It was pointed out previously that time-delay reconstruction based on measurement of a single variable $x(i)$ is only effective if this variable is dynamically linked to all of the other system variables. There is another more subtle case where the reconstruction may lose its effectiveness. This is if one of the variables is only loosely coupled to the others or in some way does not faithfully represent the dynamics. These cases can be difficult to determine *a priori*. An example can be found in the Lorenz system discussed in Chapter 1 and presented in Fig. 4.7.1. The top graph shows the attractor plotted conventionally: the x, y, and z variables are plotted along orthogonal coordinate axes. On the bottom are two time-delay reconstructions, one using the variable x and one using the variable z. It would be unfortunate if one made the choice of the z variable for the reconstruction. This is because the attractor is relatively "flat" along the z-direction, as seen in the top graph. Some advanced computational procedures have been proposed in an attempt systematically to counter such anomalous cases. In general, knowledge of the system under study, some understanding of the role of the different measured quantities, and comparison of attractor reconstructions from different system variables, can help to reduce the severity of this problem.

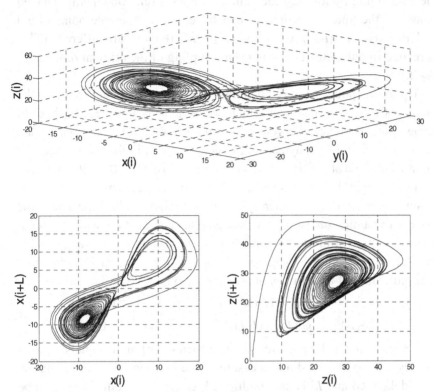

Figure 4.7.1 Lorenz attractor, formed from the three state variables in the differential equations (top), and from time-delay reconstruction using the x or the z variable (bottom). Use of z in the time-delay reconstruction does not capture the dynamics as readily as does the use of x or y.

Another way to approach this problem is to augment standard time-delay reconstruction with the use of multiple time signals. There are circumstances, such as multi-electrode EEG or EKG, where multiple simultaneous recordings arise naturally. Subsets of these multiple time series could be used in attractor reconstruction, instead of or in addition to time-delay reconstruction (e.g., Destexhe *et al.* 1988).

An unexplored possibility which might be useful in special cases is reconstruction with unequal time delays. If the time series contains dynamics on very different time scales, then selection of the time delay L is problematic. It may be possible to capture these disparate dynamics in the embedding by forming each attractor point from non-equally delayed values of the time series $x(i)$, so that for example a sample point would be [$x(i)$ $x(i-L_1)$ $x(i-L_2)$ $x(i-L_3)$ $x(i-L_4)$], where the L_i are different. (This is related to but distinct from uneven sampling of the original time series itself.)

Up to this point we have been discussing time series generated as the *output* of a nonlinear system which spontaneously advances in time. In some cases, however, the system is explicitly driven by an input signal. In order to capture the dynamics of such a system in the state space, the input signal must be included when reconstructing the attractor. An example might be the response of sensory afferents when driven by stimuli of various types; clearly the output signal tells little about the system dynamics unless it is considered in conjunction with the input signal. A straightforward formulation of this situation (Casdagli 1992) involves the creation of augmented state vectors, which include time-delayed values of both the input $u(t)$ and the output $x(t)$:

$$y(i) = [\ x(1+(i-1)J),\ x(1+(i-1)J+Lo),\ ...,\ x(1+(i-1)J+(M-1)Lo)\ ...$$
$$u(1+(i-1)J),\ u(1+(i-1)J+Li),\ ...,\ u(1+(i-1)J+(M-1)Li)\ ].$$

This approach is valid even for stochastic (random) input signals, as long as they can be observed and measured. The problem of determining the time delays Lo and Li is now multiplied, since they should in general be different for the input and the output, but extensions of such techniques as False Nearest Neighbors can be applied (Walker & Tufillaro 1999).

Another extension of the main theme of attractor reconstruction is the use of discrete event times, such as embodied in neural spike trains, as the time series $x(i)$. When system information is contained in the timing of discrete events, it is known as a *point process*. It is possible to capture the dynamics of an underlying system through appropriate manipulations on a point process generated by that system. Simply put, we can use neural inter-spike intervals (for example) to study the dynamics of the underlying (continuous-time) neural system (Sauer 1994, 1995). The validity of this approach has been demonstrated by driving a simple integrate-to-fire model neuron with a chaotic signal (e.g., Lorenz), and showing that the resulting interspike intervals can be used to reconstruct the attractor that corresponds to the signal that drives the neuron. That is, replace the $x(t)$ in the attractor reconstruction with the sequence of interspike intervals:

$$y(1) = [\, T_1 \quad T_2\text{-}T_1 \quad T_3\text{-}T_2 \quad T_4\text{-}T_3 \ ...\,]$$

$$y(i) = [\, T_i \quad T_{i+1}\text{-}T_i \quad T_{i+2}\text{-}T_{i+1} \quad T_{i+3}\text{-}T_{i+2} \ ...\,]$$

where the T_i are the times at which the spikes occur. This approach also makes feasible the study of other discrete-time phenomena such as cardiac dynamics via heart-beat intervals and oculomotor dynamics via intervals between fast-phase eye movements as alluded to above. (An important caveat is that, for some computations based on such a reconstruction, the fidelity of the reconstructed attractor in reproducing the dynamics of the underlying signal depends on the average rate at which the events are produced. Too low a spiking rate, for example, will not adequately reproduce the underlying dynamics. See section 7.7.)

Finally, we discuss an approach to noise reduction that is carried out in the state space (Kostelich & Yorke 1988, 1990). Unlike more conventional filtering, this approach does not assume time invariance or stationarity or a constant frequency spectrum; the "filtering" is not based on data points that are localized in time or frequency but rather are localized in the state space and are therefore dynamically related. The method is based on the assumption that that attractor can be described by local linear approximations: each small section of the attractor can be considered as a straight line and can therefore be modeled by a linear function. (The identical approach will be used later in the development of methods of nonlinear forecasting in Chapter 7.) To generate such local linear approximations, a reference point is selected, and its nearest neighbors in state space are used to generate a local linear fit to the attractor, via least-squares estimation in M dimensions. Then, a short trajectory segment is identified that passes near that reference point, and this segment is modified slightly so that it aligns more closely with the linear approximation. This works on the assumption that the local linear approximation, based on local spatial averaging, has better statistics than any individual trajectory path. This process is repeated for successive reference points. The entire procedure can be performed iteratively, with the modified trajectory paths replacing the original (noisy) version at each iteration. In practice, limits are placed on the maximum allowable trajectory modification so as not to alter legitimate transients, and when a given data value is modified the time-subsequent data values are also adjusted so that the time-delayed vectors still "match up" (due to time-delay reconstruction, the first element in one M-dimensional attractor point will appear as an element in another attractor point, and these should be identical). The result is an attractor (and a modified time series as well, if it is desired to recover such) that has been modified to be more "internally self-consistent" in terms of following local linear trajectories.

This is another excellent example of how clear and straightforward consideration of behavior in high-dimensional state spaces can be put to effective use in system-analysis problems. We note that although this approach has an intuitive appeal, there are more general approaches and analyses that may be useful in especially noisy situations (Casdagli *et al.* 1991).

The majority of physiological studies that have made use of reconstructed state spaces have not, to date, had to resort to such noise-reduction measures, since the signals have generally had high signal-to-noise ratios. In particular, discrete events such as neural spikes and heart beats can be preprocessed to provide very low noise levels. Even in these cases, however, noise might be present in the form of *timing jitter*, due to sampling at a fixed and practical rate. Thus even in those cases where the "noise" as conventionally considered might be low, some improvement might be obtained with noise-reduction techniques due to event times being randomly displaced.

References for Chapter 4

TA Bass (1985) The Eudaemonic Pie. New York: Houghton-Mifflin.
TA Bass (1999) The Predictors: How a Band of Maverick Physicists Used Chaos Theory to Trade Their Way to a Fortune on Wall Street. New York: Henry Holt.
M Casdagli (1992) A dynamical systems approach to modeling input-output systems. In: M Casdagli, S Eubank (eds) Nonlinear Modeling and Forecasting, SFI Studies in the Sciences of Complexity, Vol. XII. Reading: Addison-Wesley.

M Casdagli, S Eubank, JD Farmer, J Gibson (1991) State space reconstruction in the presence of noise. *Physica D* 51:52-98.

A Destexhe, JA Sepulchre, A Babloyantz (1988) A comparative study of the experimental quantification of deterministic chaos. *Physics Letters A* 132:101-106.

AM Fraser, HL Swinney (1986) Independent coordinates for strange attractors from mutual information. *Physical Review A* 33:1134-1140.

MB Kennel, R Brown, HD Abarbanel (1992) Determining embedding dimension for phase-space reconstruction using a geometrical construction. *Physical Review A* 45:3403-3411.

EJ Kostelich, JA Yorke (1988) Noise reduction in dynamical systems. *Physical Review A* 38:1649-1652.

EJ Kostelich, JA Yorke (1990) Noise reduction: finding the simplest dynamical system consistent with the data. *Physica D* 41:183-196.

C Letellier, LA Aguirre, J Maquet (2005) Relation between observability and differential embeddings for nonlinear dynamics. *Physical Review E* 71:066213-1:8.

JM Martinerie, AM Albano, AI Mees, PE Rapp (1992) Mutual information, strange attractors, and the optimal estimation of dimension. *Physical Review A* 45:7058-7064.

E Ott, T Sauer, JA Yorke (eds) (1994) Coping with Chaos: Analysis of Chaotic Data and The Exploitation of Chaotic Systems. New York: Wiley-Interscience.

N Packard, J Crutchfield, J Farmer, R Shaw (1980) Geometry from a time series. *Physical Review Letters* 45:712.

J-C Roux, RH Simoyi, HL Swinney (1983) Observation of a strange attractor. *Physica D* 8:157-266.

D Ruelle (1990) Deterministic chaos: the science and the fiction. *Proceedings of the Royal Society of London A* 427:241-248.

T Sauer (1994) Reconstruction of dynamical systems from interspike intervals. *Physical Review Letters* 72:3811-3814.

T Sauer (1995) Interspike interval embedding of chaotic signals. *Chaos* 5:127-132.

T Sauer, JA Yorke, M Casdagli (1991) Embedology. *Journal of Statistical Physics* 65:579-616.

M Shelhamer, S Zalewski (2001) A new application for time-delay reconstruction: detection of fast-phase eye movements. *Physics Letters A* 291:349-354.

F Takens (1981) Detecting strange attractors in turbulence. In: Lecture notes in mathematics, Vol.898: Dynamical systems and turbulence. Page 366. Berlin: Springer.

DM Walker, NB Tufillaro (1999) Phase space reconstruction using input-output time series data. *Physical Review E* 60:4008-4013.

H Whitney (1936) Differentiable manifolds. *Annals of Mathematics* 37:645-680.

Chapter 5

Dimensions

The subject of the dimension of state-space attractors was one of the first to be considered in detail during the rise of "chaos theory" in the 1980s. Why should we care about dimension, and what is there to it? How hard can it be, after all, simply to increase the embedding dimension during time-delay reconstruction until there are no false nearest neighbors and no trajectory crossings, and to say that the attractor has the dimension of the least embedding dimension at which this occurs?

Needless to say, there is much more to the concept and computation of dimension than this. Dimension is a critical property because it indicates how many independent state variables are required to reproduce the system dynamics in state space, and this in turn indicates how many state variables should be included in a mathematical model of the system. Aside from this practical issue, the dimension is an indicator of the degree of "complexity" of a system, and tracking any changes in dimension due to pathology or other manipulations to the system can be a useful diagnostic criterion.

To fully embrace the power and usefulness of state-space and attractor dimensions, we must expand our conceptual basis for understanding and interpreting dimensions. In particular we will be dealing with non-integer dimensions and the dimensions of fractal objects (strange attractors). In addition to allowing us to address the practical aspects of dimension for system analysis, this excursion into the world of dimensions can open up a whole new way of thinking about the world and the physical processes within it. It is a fascinating intellectual adventure in its own right.

[Two brief notes on nomenclature will be useful at this point. First, as mentioned in Chapter 4, we will use the term "attractor" to refer to a set of trajectories in state space, recognizing that an attractor is more precisely an object (*manifold*) to which the trajectories are drawn, and that not all trajectories represent attractors (they may not be stable, for example). The second point is that extensive use will be made of logarithms in this chapter, which will be expressed in most cases simply as *log*. We will take this generally to mean logarithm to the base e, or *ln*. Base 10 logarithms will be specified as log_{10}. Logarithms in the two bases are related by a scale factor: $ln(x)=log_{10}(x) \cdot ln(10) \approx 2.3 \cdot log_{10}(x)$. In many cases we will deal with the ratio of two logarithms, for which the base makes no difference since the scale factors cancel.]

5.1 Euclidean dimension and topological dimension

The dimension concept that is most familiar to us is *Euclidean dimension*. This is given by the number of independent coordinates required to specify a location on a given object (or, in the case of *Euclidean space*, the number of coordinates required to specify a location in space). Thus an idealized point in space has a dimension of zero, a line has a dimension of one, and a plane has a dimension of two. Our common experience of space (Einstein's non-Euclidean space-time notwithstanding) is that of Euclidean space with three dimensions, and the time-delay reconstructions that we met in Chapter 4 exist in M-dimensional Euclidean spaces.

A generalization of Euclidean dimension is *topological dimension*. One of the more accessible definitions (with some simplification) of topological dimension is that a point has dimension zero, and any other object has a dimension that is one greater than the dimension of a finite number of other objects that are required to cut it into separate pieces. This is a recursive definition. So, a line has dimension one, since a point of dimension zero can cut it into two pieces. A plane has dimension two, since a line of dimension one can cut it into pieces. These correspond with our intuitive sense of Euclidean dimension.

Now things get interesting. The topological dimension of a closed curve sitting on a plane (e.g., a circle) is one, not two which is its Euclidean dimension. This is because a finite number of points can cut a closed curve into pieces, the points have dimension zero, and 1=0+1. The surface of a sphere has a dimension of two (not three) since it can be cut by a line of dimension one.

Topological dimension is a *topological invariant*. As discussed in Chapter 4, this means that the dimension of an object (such as an attractor in state space) is not altered if the object undergoes a homeomorphic transformation.

5.2 Dimension as a scaling process – coastline length, Mandelbrot, fractals, Cantor, Koch

How long is the coast of Britain? This is the title of a classic paper by Mandelbrot (1967), which describes some of the basics of what would become known as *fractals* (a term that he coined). Imagine measuring a coastline with a measuring stick that is one mile in length. This gross resolution means that smaller features such as inlets will be passed over. If the measuring stick is reduced in length, more of these features will be resolved. As the measuring unit decreases in length, the apparent length of the coastline increases. While we might not be able to answer the original question, we can, motivated by this behavior, pose a different and perhaps more insightful question: does the apparent length increase in a systematic manner as the size of the measuring unit decreases, and if so in what way?

It turns out that there is an elegant mathematical law that describes this increase in apparent length. It is a form of *power-law scaling*:

$$N \propto \varepsilon^{-D} = 1/\varepsilon^{D}.$$

N is the total length, expressed as the number of units of basic length ε. As ε decreases, N increases, and the rate of increase is determined by the value of the exponent D.

If the data points for different pairs of (ε, N) are plotted on logarithmically scaled axes and connected by a best-fit line, the line is determined by the equation:

$$\log(N) = \log(K\varepsilon^{-D}) = \log(K) - D\log(\varepsilon),$$

where K is a constant of proportionality. The equation shows that a line formed by plotting $\log(N)$ as a function of $\log(\varepsilon)$ has a slope of $-D$. Thus the parameter D can be determined by this process. Values between 1 and 2 have been found for various coastlines. (This power-law scaling is of course only true over a certain range of values of ε, since ε can become larger than the coastline itself or too small to make the measurements practical. Nevertheless the scaling range can span several orders of magnitude. For examples see page 33 in Mandelbrot (1983).)

The power law reflects the fact that the coastlines in question are *self-similar*. Qualitatively this means that, given an image of a coastline with no external length reference (like a person, a plant, or a ship), one cannot determine the length scale (magnification level) of the image. A great insight came from the realization that this self-similar property is a feature of a great many physical objects and natural processes. Computer-generated self-similar images have been used to mimic trees, mountains, and lightning. Similar characteristics have been found in many anatomical objects, although we will not go into detail on this matter (Bassingthwaighte *et al.* 1994). All of these self-similar objects are example of *fractals*.

Quantitatively, self-similarity means that if the measurement unit length ε decreases by a factor of a, then the total length in terms of the number of ε-units increases by a factor of a^D:

$$\varepsilon' = \varepsilon/a \quad \rightarrow \quad N' = K\varepsilon'^{-D} = K(\varepsilon/a)^{-D} = Ka^D\varepsilon^{-D} = Na^D.$$

What makes this "self-similar" is that this property holds true no matter where we start – no matter what we choose for N and ε. This gives *power-law scaling* its unique characteristics. As a counterexample, we might consider exponential scaling, where the number of ε-sized sticks needed to cover an object is $N = Ke^{-\varepsilon}$. In this case it can be shown that a decrease in the measuring length ε by a factor a will produce different increases in N, depending on what values of ε and N we start with.

(While we will concern ourselves with measuring self-similar properties of state-space attractors, these concepts can be applied directly to time series data. Some procedures for doing this will be discussed in Chapter 13.)

The value of the exponent D in our power-law formulation is known as the *fractal dimension*. It shows, in its most general terms, how the "bulk" (length, size, mass) of an object scales as a function of the size of the reference unit used to make the measurement. Calling D a "dimension" may seem like a pure flight of fancy, but we can begin to make this believable by grounding it with some familiar cases.

Let us use the defining equation for D to find the fractal dimension of a line that has length $\mathbf{L}$. Clearly, if we measure the length of the line by comparing it to the length of a reference line of length ε, it takes $N=\mathbf{L}/\varepsilon$ of these reference lines to cover the line being measured. But:

$$N = \mathbf{L}/\varepsilon = \mathbf{L}\varepsilon^{-D} \propto \varepsilon^{-D}, \text{ if } D=1.$$

This shows plainly that the fractal dimension of a line is 1, as is its Euclidean dimension and its topological dimension. The same reasoning shows that any simple closed curve (whether it exists on a plane or in thee-dimensional Euclidean space) also has a fractal dimension of 1, which matches its topological dimension (but not its Euclidean dimension). Similar reasoning shows that the number of ε-sided squares (each with area ε^2) that are needed to cover a square with side of length $\mathbf{L}$ (area $\mathbf{L}^2$) grows as a power law with exponent 2. Thus the fractal, topological, and Euclidean dimensions of a plane are identical. A "twisted" plane, on the other hand (like a floppy sheet of paper), has fractal and topological dimensions of 2 but a Euclidean dimension of 3. A clear sign of the development of some topological intuition is if one can see how a twisted plane (or a curved line) is dimensionally unchanged from a flat plane (or a straight line).

Clearly we do not need such an odd definition of dimension just to examine lines and planes. Let's move on to some objects with more interesting scaling properties, where normal concepts of length, size, and dimension start to break down. The *middle-thirds Cantor set* is one such object. Figure 5.2.1. shows its construction. Start, at step number zero, with a line segment of length 1. In step one, remove the middle third of the line segment, leaving two segments each of length 1/3. In each subsequent step n, remove the middle third of each remaining line segment, resulting in 2^n segments each of length $(1/3)^n$ at step n. As the number of steps increases, we eventually reach a point where there is an infinite number of line segments, each of length zero, for a total length of zero. This is an infinite number of points that occupies no space.

Now let's compound the angst over such an unusual object by inquiring as to its dimension. We would like to say that the Euclidean dimension is zero, since this is just a collection of zero-dimensional points. But there is an infinite number of them, so this is not very satisfying. The topological dimension does not help much since the set is already separated into disjoint elements and so can not really be "cut" by any other object. Fractal dimension provides a way to capture the bizarre nature of this creation.

Referring back to the definition of fractal dimension as the exponent D in a power-law scaling process, we see that it fits precisely this situation. Here N is the number of segments and the length of each segment corresponds to ε, the length of a line needed to measure those segments. We set N in the definition equal to the number of segments created in the construction at step n:

$$N = 2^n = [(1/3)^n]^{-D} = (1/3)^{-nD}$$
$$2 = (1/3)^{-D}$$
$$\log(2) = -D\log(1/3) = D\log(3)$$
$$D = \log(2)/\log(3) \approx 0.6309$$

The fractal dimension turns out to have a fractional value. While this is a new development in terms of dimension values, it captures an intuitive sense that this infinite set of points with zero length is "more than a point" but "not quite a line." Thus the dimension falls between that of a point and that of a line.

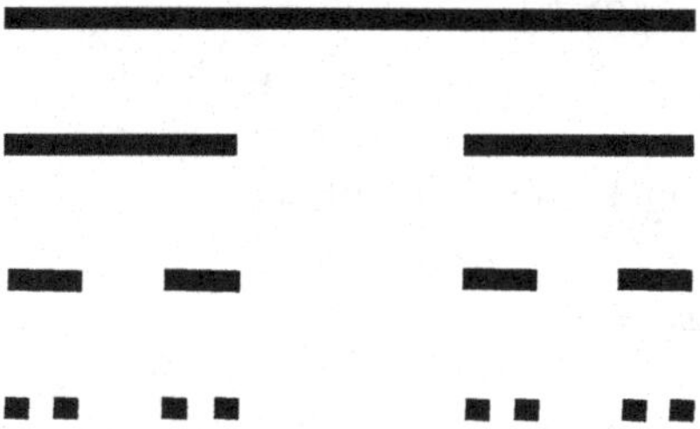

Dimensions

	step (n)	number of segments	segment length	total length
▬▬▬▬▬▬▬▬▬▬	0	1	1	1
▬▬▬▬ ▬▬▬▬	1	2	1/3	2/3
▬▬ ▬▬ ▬▬ ▬▬	2	4	1/9	4/9
▪▪ ▪▪ ▪▪ ▪▪	3	8	1/27	8/27
	N	2^N	$(1/3)^N$	$(2/3)^N$
	∞	∞	0	0

Figure 5.2.1. Construction of the middle-thirds Cantor set, or "Cantor dust," which has infinitely many points but zero length.

There are many such self-similar fractal objects that can be generated with simple recursive rules. We will study one more before moving on to state-space attractors. Construction of the *Koch curve* is demonstrated in Figure 5.2.2. Begin with an equilateral triangle with sides of unit length, in step zero. In step one, remove the middle third of each side of the triangle and tack on a triangle with a side length of one-third the original. Repeat this procedure without bound. As the figure shows, the "curve" takes on finer and finer detail as the construction proceeds. It is truly self-similar, because if one small section is magnified (no matter how much), it will look like exactly like a larger section. The total length of the curve – the sum of the lengths of all of the tiny triangles – approaches infinity. Yet, the entire object can be circumscribed by a circle with a radius of less than one, and therefore the area is finite. It has infinite length yet is bounded in space. This gives a sense of how strange a fractal can be.

The dimension can be found as in the case of the Cantor set, by setting N in the dimension definition equal to the number of segments in the construction:

$$N = 3 \cdot 4^n = [(1/3)^n]^{-D} = (1/3)^{-nD}$$

$$3 \cdot 4^n = (1/3)^{-nD}$$

$$\log(3) + n\log(4) = -nD\log(1/3) = nD\log(3)$$

$$D = \frac{\log(3) + n\log(4)}{n\log(3)} \quad as \ n \to \infty$$

$$D = \lim_{n \to \infty} \frac{\log(3) + n\log(4)}{n\log(3)} = \frac{\log(4)}{\log(3)} \approx 1.2619$$

Here use is made of the fact that the scaling is evident for small segment sizes, which means large values of the step size n. The dimension is between the topological dimensions of a line and a plane, again reflecting the intuitive sense that a line of infinite length is "more than a line" but "not quite a plane."

We will not see such clear-cut examples as these mathematical constructions in our work on actual attractors generated from experimental data. Nevertheless, we will meet attractors that can be considered as fractal objects, with non-integer fractal dimensions. Such attractors can arise from systems with chaotic dynamics, and are termed *strange attractors*. They occupy a well-defined and bounded region of the state space. Yet the system behavior is aperiodic, so no matter how much data we acquire the attractor trajectory will never return to the same location in state space (within the limits of measurement resolution and smearing due to noise). How can we jam a (potentially) infinitely long trajectory, which never repeats or crosses itself, into a finite volume of space? One way to accomplish this is if the attractor forms a fractal, such that there is finer and finer detail as we look at it more and more closely; in this sense, no matter how "dense" the trajectory in any given area of the state space, there is always room to squeeze in another trajectory passage. (In fact, some early studies equated the finding of non-integer dimension with the presence of chaotic dynamics, often erroneously as we shall see.)

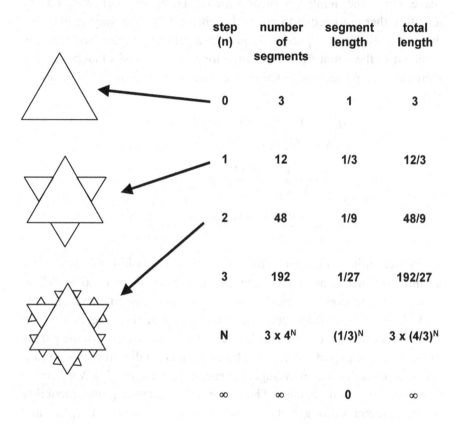

step (n)	number of segments	segment length	total length
0	3	1	3
1	12	1/3	12/3
2	48	1/9	48/9
3	192	1/27	192/27
N	3×4^N	$(1/3)^N$	$3 \times (4/3)^N$
∞	∞	0	∞

Figure 5.2.2. Construction of the Koch curve, or "Koch snowflake," which has infinite length but finite area.

5.3 Box-counting dimension and correlation dimension

Given this new concept of dimension – fractal dimension as a scaling process – our job now is to apply it to determine the dimension of an attractor in state space. We will follow the reasoning above, where dimension is defined as the exponent in a power-law scaling process.

The *box-counting dimension* implements the idea of power-law scaling in a more general form. Given an object in an M-dimensional space, count the number N of M-dimensional boxes, each with side of length ε, that are needed to cover the object. (Another way to envision this is that the entire space is filled with a grid of ε-boxes, and a box is included in the count $N(\varepsilon)$ if the attractor trajectory visits that box.) If N increases as a power law function of ε, then we can define the dimension D_B:

$$N(\varepsilon) \propto (1/\varepsilon)^D \quad as \quad \varepsilon \to 0$$

$$\log(N) = \log(k) + D\log(1/\varepsilon) \quad as \quad \varepsilon \to 0$$

$$D = \frac{\log(N) - \log(k)}{\log(1/\varepsilon)} \quad as \quad \varepsilon \to 0$$

$$D = \lim_{\varepsilon \to 0} \frac{\log(N) - \log(k)}{\log(1/\varepsilon)} = \lim_{\varepsilon \to 0} \frac{\log(N)}{\log(1/\varepsilon)}$$

Box-counting dimension has a clear relationship to power-law scaling, and hence an intuitive appeal. However, there is some debate both as to the computational feasibility of box-counting and also its usefulness in truthfully characterizing the spatial properties of an attractor (Greenside *et al.* 1982, Molteno 1993, are just two of the papers in this debate). In particular, as the box size gets smaller, fewer points are enclosed in each box, on average, yet each box is included in N no matter how many points it contains. Thus information regarding the probability of the attractor visiting certain boxes is lost. For these conceptual and practical reasons, box-counting dimension has been almost completely surpassed in most applications by the *correlation dimension*. (The reader should feel free to consider using the box-counting dimension, as it has its adherents. While computational details given here will not be directly applicable, the main points and applications will apply to both dimension estimates.)

Box-counting dimension is one of a series of dimensions based on a more general form. The general form allows more or less weight to be placed on how often different locations in the state space are visited by the trajectories. The derivation of these dimensions will be outlined here, but this information is not needed in order to make use of dimension measurements in practical applications, and so the reader is invited to skip ahead to the next section if these mathematical details are of no interest.

The most general form of these *Renyi dimensions* is:

$$D_q = \frac{1}{q-1} \lim_{\varepsilon \to 0} \frac{\log I(q,\varepsilon)}{\log(\varepsilon)}.$$

Here, q indicates which in the series of dimensions is being considered ($q=0$ is the box-counting dimension), ε is the size of a box as before, and

$$I(q,\varepsilon) = \sum_{i=1}^{M(\varepsilon)} [\mu(C_i)]^q$$

$$\mu(C_i) = \lim_{T \to \infty} \frac{\eta(C_i,T)}{T}$$

The quantity $\eta(C_i,T)$ is the amount of time that a trajectory spends in box C_i in the time span from 0 to T. Hence, $\mu(C_i)$ is the proportion of time that the trajectory spends in box C_i (in the long run, as T increases), and this is essentially the probability that the attractor trajectory passes through box C_i.

If $q=0$, note that $\mu(C_i)$ is raised to the power zero in the equation for $I(q,\varepsilon)$, so that $[\mu(C_i)]^q$ is zero if C_i is not visited at all, and one if it is visited by the trajectory no matter how briefly. In other words, $I(q,\varepsilon)$ is a count of the number of boxes C_i visited by the trajectory. Noting the similarity of the equation for D_q to that above for the box-counting dimension (and noting that $\log(1/\varepsilon) = -\log(\varepsilon)$), it is clear that D_0 is indeed the box-counting dimension.

Incrementing q to 1, the next in the series of dimensions, D_1, is known as the *information dimension*. Although we will omit the details, the name derives from the fact that the expression for D_1 resembles that for the information content of a data set, in the sense of Shannon information theory (Shannon & Weaver 1963).

The next dimension, when $q=2$, is of by far the most interest to us. It is called the *correlation dimension*, and the equations above reduce to:

$$D_2 = \lim_{\varepsilon \to 0} \frac{\log\left[\sum_i \mu^2(C_i)\right]}{\log(\varepsilon)}$$

In a paper that has become a true classic (cited by many, read by few), Grassberger and Procaccia (1983) showed that the summation in this equation could be approximated by a *correlation integral* which is much easier to compute from experimental data:

$$\sum_i \mu^2(C_i) \cong \frac{1}{N^2} \sum_i \sum_j \mathbf{U}(\varepsilon, |y_i - y_j|) \quad (i \neq j)$$

$$\mathbf{U}(\varepsilon, |y_i - y_j|) = \begin{cases} 1 & \text{if } |y_i - y_j| < \varepsilon \\ 0 & \text{otherwise} \end{cases}$$

Although expressed as a discrete summation, the quantity on the right in the upper equation is known as a correlation integral. The operator $\mathbf{U}(\cdot)$ is a step function; as expressed here, it is one if the distance between the attractor points y_i and y_j are within distance ε of each other, and zero otherwise. Thus, the correlation integral counts the number of pairs of points on the entire attractor that are within distance ε of each other, and divides this by N^2, the total number of pairs of points.

The demonstration of this equality can be found in Grassberger and Procaccia (1983), but an intuitive argument can be made to justify it. If, at a given box size ε, the (discretized) attractor visits box C_i for P points out of a total number N of points on the attractor, then $\mu^2(C_i)=(P/N)^2$. On the other hand, in box C_i, since there are P points there will be approximately P^2 pairs of points – that is, box C_i will contain P^2 pairs of points within distance ε of each other. By the definition of the function U, this means that C_i will contribute an amount P^2 to the correlation integral. Since this is divided by the total number of point pairs N^2, this contribution $((P/N)^2)$ is identical to that of the contribution of C_i to the summation of $\mu^2(C_i)$, and the two quantities are equal. Actually, the equality is an approximation, largely due to the fact that the correlation integral is expressed in terms of inter-point distances and therefore implies a "ball" of radius ε to establish the criterion distance ε, while the original definition of the dimension is based on a cube with side length of ε. For most practical applications the approximation is close, and improves as N increases.

Since it is so easy to compute, the correlation dimension has become a standard measure of the fractal dimension of attractors that have been reconstructed in the state space. It approximates, and is a lower bound for, the box-counting dimension (i.e., it is less than or equal to the box-counting dimension, with equality in the case when all the boxes C_i are occupied equally). Its use is simple in principle, but nontrivial in practice. We discuss the practicalities next.

5.4 Correlation dimension – how to measure it correctly

Now we turn our attention to the practical problem of measuring the fractal dimension of an attractor. To review, recall that we have determined that the correlation integral can be used to approximate the correlation dimension:

$$D_2 = \lim_{r \to 0} \frac{\log[C(r)]}{\log(r)}$$

$$C(r) = \frac{1}{N(N-1)} \sum_i \sum_j U(r, |y_i - y_j|) \quad (i \neq j)$$

The notation has been changed to use r (radius) rather than ε to designate the criterion distance; when two points y_i and y_j are closer together than r, they are "spatially correlated" and contribute to the correlation integral (actually summation) $C(r)$. The divisor has also been changed to reflect the fact that, since the case $i=j$ is always skipped in the summation (since the distance between y_i and y_j is zero when $i=j$ and counting this does not accurately reflect how close different points are to each other), the total number of inter-point pairs being compared is $N(N-1)$ rather than N^2.

If $C(r)$ increases as a power-law function of r, then $C(r)$ versus r on a log-log plot should be a straight line, and the slope will be the correlation dimension D_2. The construction of $C(r)$ is even simpler than suggested by the equation:

1. Reconstruct the attractor in an M-dimensional embedding space as described in Chapter 4.

2. Choose a reference distance r.
3. Select a *reference point* y_i on the attractor. (This corresponds to summation with index i held constant.)
4. For every other point y_j, find the distance between this point and the reference point y_i, and if this distance is less than r, add one to the correlation integral. (This corresponds to summation over index j.)
5. Choose the next point as a reference (increment i), and repeat step 4, comparing all points to this new reference point.
6. Divide the summation, the accumulated number of point pairs that are closer than r, by $N(N-1)$.
7. Repeat steps 3-5 for another reference distance r.
8. Plot $\log[C(r)]$ versus $\log(r)$. The slope is the dimension D_2 (or D_{corr}).

The astute reader may note a discrepancy here between theory and practice. The definitions of dimension involve a limiting process, specifying power-law scaling as the criterion distance r or ε decreases. Yet the correlation dimension is determined over a *scaling region* which does not necessarily include the smallest available values of distance r. The correct way to think of the limiting process is that it specifies the existence of power-law scaling over a range of distances, and only in the idealized case of a mathematical construction will the scaling hold for infinitesimally small distances.

One immediate question is what range of reference distances (r) to use. Obviously one can increase r until all pairs of points are included and $C(r)=1.0$. If an estimate can be made of the minimum inter-point distance, then it can be used as the smallest value of r.

An example of correlation integrals from analysis of the Lorenz system is shown in Figure 5.4.1, in the top graph. The attractor was reconstructed from 8000 values with time-delay embedding, using embedding dimensions of 5 to 10. In the graphs, there is a line for each of the six embedding dimensions, although they overlap almost entirely. As shown in the figure, the correlation integrals $C(r)$ have their maximum value when the criterion distance r is large enough to include all pairs of points; $C(r_{max})=1$ and so $\log[C(r_{max})]=0$. Power-law scaling is evident as straight lines on this log-log plot, over a range of $\log_{10}(r)$ from approximately −0.5 to 1.0.

The smaller vertical lines near the left end of the graph give an indication of the smallest inter-point distances. For each correlation integral $C(r)$ the minimum inter-point distance ($|y(i)-y(j)|$) was found for each reference point $y(i)$, and the mean of these minima (across all reference points) was determined and plotted as a vertical marker. These markers indicate the distance r below which noise rather than attractor dynamics dominates the computation of $C(r)$. Computing and graphing this information provides a diagnostic criterion when looking for scaling: any power-law scaling at r values below these is suspect. In the example here, it is clear that scaling breaks down before this lower limit is reached; this is not always so clear.

The bottom graph in the same figure shows the slopes of the correlation integrals. The slopes are approximately constant over the range given by the thick horizontal line; this is known as the *scaling region*. The slope over this scaling region, averaged over all six values of M, is 2.05, which matches well the established dimension of this system.

Several useful guidelines have been suggested to make these computations of D_{corr} less subjective (Albano *et al.* 1988). There should be less than 10% variation in the slope of $C(r)$ across the scaling region. There are different ways to interpret this; one approach is to expand the scaling region in small increments of r while any incremental changes in the slope are less than 10%, and take as the scaling region the largest such range of r. Ideally the scaling region should be at least one log unit in length (one factor of 10). When the scaling region has been identified from the slope graph, go back to the graphs of $C(r)$ and find the slope over the scaling region with a linear regression. Finally, there should be less than 10% variation across consecutive values of M.

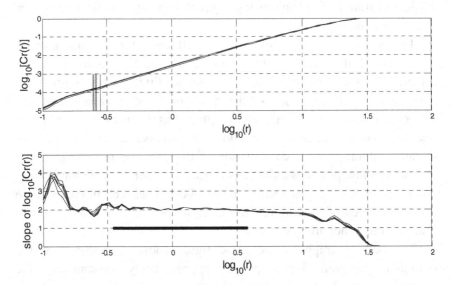

Figure 5.4.1. Correlation integrals (top) and their slopes (bottom), as steps in the determination of the correlation dimension of the Lorenz attractor.

The embedding dimension in the case examined here was verified by finding the correlation dimension with different values of M until D_{corr} no longer increased with M, as shown in Figure 5.4.2. The false nearest neighbors method could be used as well.

It is worth noting that although an embedding dimension of at least twice the attractor dimension is required to guarantee a proper embedding (reconstruction), in fact it has been shown (Ding *et al.* 1993) that an embedding dimension that is *at least equal to* the attractor dimension is sufficient for reliable computation of the correlation dimension. This has been shown mathematically (theoretically) and observed numerically, and holds for large data sets. For shorter data sets (just a few thousand points), larger embedding dimensions may be required before D_{corr} reaches a plateau with M, although not necessarily as large as two times the attractor dimension.

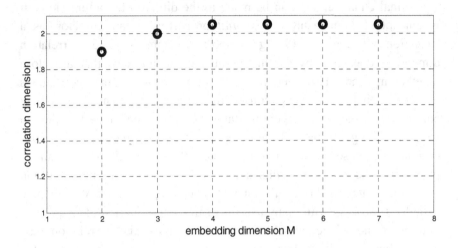

Figure 5.4.2. Saturation of the correlation dimension with increasing embedding dimension.

The reconstruction time delay L was initially set based on the correlation time as discussed in Chapter 4. Correlation time for this system is 15 (in arbitrary units of sampled time), and so $L=30$ is a reasonable starting point. However, in practice these computations made use of a recommendation of Albano *et al.* (1988). Instead of determining a value for L and keeping it constant as M is varied, an *embedding window* T_w is established, and this is kept constant across changes in M. It is defined as $T_w=[M-1]L$. Recall from Chapter 4 that each point on the attractor is given by: $y(i)=[x(i) \; x(i+L) \; x(i+2L) \; ... \; x(i+(M-1)L)]$. The time spanned by the time-series values $x(i)$ that make up a single attractor point $y(i)$ is thus $[M-1]L$, or T_w. The duration of the window T_w can be established based on the same reasoning as originally used for L: the window should span a time period that is long enough so that the elements of $y(i)$ are not too close in time and therefore redundant, and short enough that they approximate a time derivative in the sense of capturing the flow of the trajectory. As recommended for L, a suitable value is some small multiple of the correlation time. As M is increased in the course of the dimension estimation, L should be reduced in order to keep T_w approximately constant: $L=T_w/[M-1]$.

A final enhancement can be made to the dimension computations. It is our intent to measure the *spatial* properties of an attractor, as a reflection of the underlying temporal dynamics. The correlation dimension does this by finding points on the attractor that are close together in space. However, the attractor is created from a continuous trajectory that represents consecutive points in time, and these consecutive time points will in many cases also be close together in space along a given path of the trajectory. This can give rise to spurious correlations based on temporal rather than spatial or "dynamic" proximity. The way around this is to exclude from the computation of the correlation integral those pairs of points that are close together in time (Theiler 1986). As with many aspects of the determination of D_{corr}, there is both art and science to selecting the duration of this "correlation dead zone," but a value close to the correlation time (time for the autocorrelation function to decay by $1/e$, see Chapter 4) is a reasonable place to start.

As if these guidelines and precautions were not enough, there is yet another safeguard that should be applied after a D_{corr} estimate has been made (Eckmann & Ruelle 1992). Assume that the attractor has N points, with a maximum extent of D in M-space. Let $\mathcal{N}(r)$ be the number of unique pairs of attractor points that are within r of each other (up to this point we have been counting the distance between $y(i)$ and $y(j)$ and the distance between $y(j)$ and $y(i)$ separately, here we consider them the same and count only one). So, when the distance r is large enough all point pairs are counted: $\mathcal{N}(D) = N^2/2$. We are assuming that power-law scaling holds, so that $\mathcal{N}(r) = kr^d$. Then:

$$\mathcal{N}(D) = N^2/2 = kD^d \quad \text{and} \quad k = N^2/2D^d$$

which means that

$$\mathcal{N}(r) = kr^d = (N^2/2D^d)r^d = (N^2/2)(r/D)^d.$$

For sufficient statistics the sum $\mathcal{N}(r)$ should have "many" points:

$$\mathcal{N}(r) \gg 1 \quad \Rightarrow \quad (N^2/2)(r/D)^d \gg 1$$

Substituting into this inequality the quantity $\rho = r/D$:

$$\mathcal{N}(r) = (N^2/2)(r/D)^d = (N^2/2)\rho^d \gg 1$$
$$N^2 \rho^d \gg 2$$
$$2 \log N + d \log \rho \gg \log 2$$
$$2 \log N \gg \log 2 - d \log \rho$$
$$2 \log N \gg d \log \rho^{-1}$$
$$d_{max} = \frac{2 \log N}{\log(1/\rho)}$$

This sets a bound on the maximum value of D_{corr}, which is d_{max}. Recall that the quantity ρ was defined above as r/D; it is, in words, the proportion of the total attractor size (maximum extent) over which power-law scaling is to be found. Clearly ρ cannot be too large or we will start to run out of attractor points in the summation. A reasonable choice for ρ is 1/10; that is, we do not expect to find scaling once we have made the criterion distance r more than 10% of the overall size of the attractor. (Note that even in the case of the well-behaved noise-free Lorenz attractor in Figure 5.4.1, the scaling region extends only to within about one $\log_{10}$ unit, or factor of 10, of the maximum value of r.) This leads to:

$$d_{max} = \frac{2 \log N}{\log(1/0.1)} = \frac{2 \log N}{\log(10)} = 2 \log N$$

For $N=1000$ this means that $d_{max}=6$, and for $N=100,000$ we find that $d_{max}=10$. This limit provides a quick sanity check on values of D_{corr}. (Note: there are published values that violate this bound, especially in the earlier papers.)

Following these steps will help to assure that accurate and reliable dimension values have been obtained. Still, there are a few more guidelines for validating dimension estimates. D_{corr} should be relatively robust to variations in T_w, to changes in N, and to changes in the sample rate of the underlying signal.

In summary, we can provide final a set of **guidelines for the estimation of the correlation dimension**:

1. Set the embedding window T_w (possibly based on consideration of the correlation time).

2. Find D_{corr} using the method above, for different values of M, changing L in order to maintain T_w constant.
3. Incorporate the dead-zone window to avoid spurious temporal correlations.
4. Find the scaling region for each correlation integral based on a threshold of 10% variation in slope as the range of r is extended.
5. Having identified the scaling region from the graphs of the slopes of $\log[C(r)]$ vs. $\log(r)$, perform a linear regression on the correlation integrals $C(r)$ themselves over this scaling region in order to get the slope ($\sim D_{corr}$) for a given set of parameters.
6. Verify that D_{corr} does not change by more than 10% over several consecutive values of M, while verifying that D_{corr} has saturated with increasing M.
7. Verify that the Eckmann-Ruelle limit of D_{corr} based on N has not been exceeded.
8. Finally, assess the robustness of D_{corr} with respect to changes in N, T_w, and other parameters. Variations of 10% of more should be examined, depending on the particular situation.

One of the stickiest issues is the establishment of "error bars" on dimension estimates. We cover this in the next section, although a different approach to establishing the reliability and aiding in the interpretation of D_{corr} is provided by *surrogate data* techniques, covered in Chapter 6.

5.5 Error bars on dimension estimates

The problem with assigning confidence intervals (error bars) to dimension estimates is that of error propagation. Error estimates in linear regression, for example, can be obtained because the data manipulations in performing the regression are fully known and straightforward. If it is assumed that measurement noise has a Gaussian distribution, the mean and standard deviation of the data can be carried through the calculations to yield estimates of the statistical deviation of the resulting regression parameters. The situation with dimension calculations is not so simple, because the data manipulations are complex and a model for the data is not generally available (otherwise there would be little use in performing dimension computations).

Although there are methods available to make such error estimates, they are not widely used in practice. Instead, dimensions are generally compared across different conditions to look for changes, and the powerful techniques of surrogate data are used (see Chapter 6).

Perhaps the most general theoretical approach to the problem is that taken by Theiler (1990), who showed that in many cases we can expect the variation in the dimension estimates to decrease as $N^{-0.5}$. Beyond that, it is difficult to determine much in the way of general rules since the variability depends on such details as the distribution of the points in space, the correlation time, and other particulars of the attractor.

A more empirical approach (Ramsey & Yuan 1990) made use of extensive simulations carried out for several example systems. Variation in D_{corr} estimates is due to both small sample sizes and high embedding dimensions: "the relative sparseness of the data at high embedding dimensions increases the sensitivity of the slope coefficient estimates to relatively small variations in the sampled spatial distribution of the data." This study presented equations for the mean and standard deviation of the D_{corr} estimates as functions of N and M, and verified them with a number of standard examples. The equations are difficult to apply directly because their parameters depend very strongly on the properties of each particular attractor. However, the authors suggest an approach that might make the equations useful in practice. First draw small sub-samples from the overall signal, and find D_{corr} for each of these sub-samples. Repeat this for different sub-sample sizes N and different embedding dimensions M (although the sub-samples for a given N should not overlap, so that the D_{corr} estimates are statistically independent). From this empirical data, determine the mean and standard deviation of the D_{corr} estimates at different values of N and M, and use these data to fit the parameters of the equation that describes the impact of N and M on the mean and s.d. of the D_{corr} estimates. Then, with the parameters determined, use these equations to estimate the mean and s.d. of the D_{corr} estimate for the full data set of maximum length N.

The equation for the mean is rather involved, but the one for standard deviation is more approachable:

$$\log(\text{s.d.}) = \log(\sigma_D) = \alpha_1 + \alpha_2 \log(N) + \alpha_3 \log(M) + \alpha_4 \, M/N.$$

Since α_4 is generally smaller than the other parameters, and M/N is small for typically large values of N, this equation is approximately:

$$\text{s.d.} = \sigma_D \propto N^{\alpha_2} M^{\alpha_3}$$

which shows more clearly the relations involved. The value of α_2 is on the order of -0.5 for several systems that were examined – s.d. decreases in proportion to the square root of sample size N. Values of α_3, the variation with embedding dimension M, are much more dependent on the specific system.

5.6 Interpretation of the dimension

Interpretation of dimension values depends on what one is trying to demonstrate about the system in question.

Many early studies applied the correlation dimension in an attempt to "prove" that the underlying system exhibits chaotic behavior. If the attractor had a fractional dimension, it was a strange attractor, and must have come from a chaotic system. The trick here – and it is not trivial – is to determine, under the limitations of limited noisy data, when the dimension is truly not an integer. Given the fact that dimension estimates can depend on the specific computational parameters, that sufficient care was not always taken in showing robustness with respect to these parameters, and that there is no accepted means of setting confidence intervals on the dimension values, this was a highly suspect endeavor. Since the time of these early studies it has become much harder to demonstrate convincingly that a system is chaotic.

This situation was exacerbated with the recognition that filtered noise can exhibit a finite correlation dimension, compatible with that from a chaotic system (Rapp 1993, Rapp *et al.* 1993). Purely random processes with power-law frequency spectra (spectra that decay as $1/f^{\alpha}$) can also yield finite dimensions (Osborne & Provenzale 1989). These are especially troubling because *noise is infinite dimensional*, and ideally the correlation dimension should increase without limit with increasing embedding dimension. Clearly by the early 1990s it was time to take a step back and gain some perspective on what can be claimed based on dimension values. (As Paul Rapp put it at a conference at that time: "I am going to have a busy career in the 1990s correcting the mistakes that I made in the 1980s" – or words to that effect.)

Even when the "chaoticity" or dimension cannot be determined with certainty, there are cases in which relative dimension values – dimensions obtained under different stimulus conditions or health versus pathology – can be useful. (Rapp 1993). Also, by seeing how the dimension changes under certain data manipulations (surrogate data, Chapter 6), one can use it as a tool to examine the degree of nonlinearity and randomness in a system, among other properties.

One of the more straightforward interpretations of the dimension is as the number of state variables needed to capture the dynamics of the system. This is a useful way to think of dimension, but it has practical limitations because even knowing the number of state variables one cannot generally work back to the equations that describe the system (see Chapter 14 for some attempts in this direction). There are also cases in which signals generated by horrendously large and complicated systems produce almost absurdly small dimensions. Some early studies on human EEG show this property, leading one to question what dimension means in a case like this. (Some of these studies did, however, violate some of the guidelines given here for reliable dimension estimates, casting some doubt on the numerical values reported.) Thus it seems that using the dimension to indicate a more general sense of "dynamic complexity" may be more fruitful.

An apparent paradox can arise in attempting these interpretations. Consider for example a very simple linear dynamic system: a simple harmonic oscillator as discussed in Chapter 4, which can be described by the differential equation:

$$\ddot{x} = -\omega^2 x$$

or equivalently by the pair of equations:

$$\dot{x}_1 = \omega \, x_2$$
$$\dot{x}_2 = -\omega \, x_1$$

What is the dimension this system? One can argue that it is either one or two. The first equation above shows that only a single state variable x is needed to describe the dynamics, which suggests that the dimension is one. However, two initial conditions are needed for this equation (specifying initial position and velocity), suggesting that the dimension is two, as does the fact that two first-order equations are needed. In the state space, as we know from Chapter 4, the trajectory forms a closed loop or ellipse, which has a topological dimension of one but a Euclidean dimension of two, and the correlation dimension is one as well. But it needs to be embedded in $M=2$ dimensions in order to avoid trajectory crossings. Is there a resolution?

Like any good paradox, the value of this one is mainly in getting us to think more deeply about the meaning and implications of the underlying concepts. Nevertheless a case can be made that this system has one-dimensional dynamics. In the state space, once the trajectory is established, in fact only a single *phase* variable is needed to specify the current state of the system. A simple model for such an oscillator is $\dot{\theta}$=constant, which clearly involves only a single variable, the knowledge of which is sufficient to locate the current state of the system on the trajectory. The fact that the trajectory has to be embedded in two-dimensional space (a plane) in order to avoid self-crossings simply reflects the fact that the embedding theorems stipulate an embedding dimension of at least twice the attractor dimension for a proper embedding (see Chapter 4). (The author is grateful to Dan Kaplan for helping to provide these insights, among many others. In turn I heartily recommend his 1995 book as a general introduction to nonlinear dynamics.)

5.7 Tracking dimension over time

The dimension is a global property of an attractor; it carries no temporal information. There are, however, cases in which it would be helpful to track the "instantaneous" dimension of a system as its attractor evolves.

The obvious way to do this is to compute a separate correlation integral $C(r)$ for each reference point ($y(i)$ in the defining equations above), instead of averaging over all reference points as is normally done in order to obtain better statistics. This has been called (Farmer *et al.* 1983) the *pointwise-D_2* (or pointwise-D_{corr} in our nomenclature). In this way, although all points $y(j)$ in the state space are compared to the reference point, it is possible to assign a specific dimension estimate to that reference point, and therefore to the point in time that is represented by that reference point. (Of course, due to time-delay reconstruction each M-dimensional reference point consists of M values from the time series $x(i)$, but it is typical to assign as the "time" of the point $y(i)$ the time of the first element $x(i)$, that is, the time represented by index i.)

A refinement has been suggested (Skinner *et al.* 1993), called *point-D_2* or *PD2*. In this modification, only those reference vectors that lead to valid dimension estimates are used. This is established by two criteria. First, the correlation integral must exhibit power-law scaling for that reference point. Second, the dimension derived from the slope of the correlation integral must converge to a constant value as embedding dimension M is increased. This method has been used to track "instantaneous" changes in the dimension of heartbeat intervals as a predictor of imminent ventricular fibrillation (Chapter 19).

5.8 Examples

Good examples of the application of correlation dimension to physiology are difficult to present at this point, since the best and most recent studies make extensive use of surrogate data techniques, which we will meet in the next chapter. In fact it is difficult or impossible now to publish a research paper on the correlation dimension of an experimental system without also presenting corroborating surrogate data. We can, however, discuss a few early examples, somewhat for their historical interest but also because of the clarity of their presentations.

Among the earliest efforts to apply the correlation dimension to the study of human brain dynamics is that of Babloyantz & Destexhe (1986), which looked at dimensions during petit mal epileptic seizures. EEG samples recorded while subjects were awake, or during REM sleep, were of indeterminate dimension: the correlation dimension did not reach a plateau as embedding dimension increased to 10. This suggests that these brain states, at least as manifest in the surface EEG, have dimensions greater than 5 or are random. Stage 2 sleep, on the other hand, produced a dimension of about 5.03, and stage 4 sleep, 4.05. Epileptic EEG yielded the lowest dimension of all, 2.05. The low dimensions of these EEG recordings calls into question the interpretation in terms of the number of state variables needed to model the system. With billions of neurons interconnected in trillions of ways, what does it mean for the EEG to have a dimension on the order of 2 to 5? While this question remains open, one conclusion is clear and has continued to find experimental support: there is often a significant decrease in dimensionality in pathological conditions. As discussed in Chapter 1, these conditions might represent an unhealthy reduction in system flexibility and information-processing capability, reflected in low dimensions.

These same authors went on to examine the dynamics of cardiac inter-beat intervals (Babloyantz & Destexhe 1988), a topic which we cover later in some detail in Chapter 19 on cardiac dynamics. The *PD2* algorithm for the detection of imminent fibrillation will be discussed there as well.

While we will examine examples of neural dynamics in later chapters, one case here will serve to illustrate the state of the art of the initial round of research in this area (Rapp *et al.* 1989). EEG was recorded from human subjects while resting, performing mental additions by two, and performing mental subtractions by seven. Dimensions were computed from two-second EEG segments, sampled at 500 Hz ($N=1000$, a typical lower limit for such studies). Although the high variability of the dimension estimates was emphasized, representative values were provided which show that the resting condition has the lowest dimension, with mental addition and subtraction of higher dimensionality (medians: 3.4, 4.8, 4.8, respectively). Resting

EEG was recorded between each arithmetic segment; the dimensions in these segments did not return to the original resting value obtained at the start of the recording session. This suggests that some remnant of the preceding mental activity may be retained in the intervening resting EEG, or it may simply be the result of nonstationarity even in the resting condition. The high dimension values, especially in the arithmetic conditions, are surely cause for concern, especially given the small sizes of the data sets. Although the careful application of the guidelines given here was stressed, also presented was the case for comparative measurements of dimension in different situations. Some relaxation of the strict computational guidelines was suggested in this case, with the understanding that the resulting individual dimension estimates cannot be considered definitive. Although it must be questioned what exactly is being measured in this case by the dimension computations (Rapp 1993), the ability to associate different mental states with a gross measure such as dimension, and the resulting ability to distinguish between these mental states, holds great promise for research and diagnosis.

5.9 Points for further consideration

Filtering is ubiquitous in data-acquisition and signal-processing applications. As mentioned above, filtered noise can yield a finite non-integer dimension, which can be misinterpreted as arising from a chaotic system. Furthermore, low-pass filtering a chaotic signal can change its apparent dimension (Mitschke *et al.* 1988). Although low-pass filtering will make the signal "smoother" and therefore apparently "less chaotic," in fact such filtering can *increase* the dimension by adding a new state variable to the system, depending on the cutoff frequency of the filter. This behavior has been verified in the case of reflexive eye movement waveforms (Shelhamer 1997; see also section 15.1 of Chapter 5).

A classic conundrum in the analysis of physiological signals involves potential *nonstationarity*. While there is a specific technical description (we touched on this in Chapter 3), for our purposes "nonstationarity" refers to a change in the properties of a system over time. Obviously all physiological systems are nonstationarity, for such is life. However, in many cases it is possible to make measurements over a short enough time span with stimulus and environment held constant, so that the system can be considered to be operating in a steady state. Limiting the data acquisition process to this time period may allow an investigator to consider the system to be stationary. On the other hand, any measurements that might be made on the acquired signals will benefit from increased data-set sizes. Thus there is a tradeoff between the desire to obtain as much data as possible and the desire to treat the system as being in a steady state (these points were raised in Chapter 1). There is no clear general solution to this matter except for knowledge of the system under study, proper experimental procedure, and judicious application of the computational procedures. One example is to segment the data record and obtain independent dimension estimates from different epochs and see if they change with time.

Given the desire to obtain as much data as possible, one might be tempted simply to increase the sampling rate F_s of the time series $x(t)$. Clearly there is a lower bound on F_s, established by the Shannon sampling theorem and the Nyquist sampling rate (at least twice the highest frequency contained in the signal), but is there an upper bound? It turns out that in fact you do not get something for nothing, and too high a sample rate can increase the computational burden (more data to process) with no scientific return. Recall, for example, the "temporal dead-zone" described above for omitting points from the correlation integral that are temporally but not spatially correlated. Oversampling exacerbates this effect, leading in some cases to spurious low-dimensional correlations that can impinge on the true scaling region and make it shorter and hence less reliable. (A thorough study of the effect of sample rate, number of points, time span of the data, and filtering, has been carried out for dimensional analysis of eye movements (Shelhamer 1997).)

The nomenclature for attractor reconstruction (Chapter 4) and many of the guidelines for computing the correlation dimension (earlier in this chapter) are drawn largely from the work of Albano *et al.* (1988). In this same paper, the authors introduce a method for improving the computation of the correlation dimension, which makes use of *singular value decomposition* (SVD), an idea put forward earlier by Broomhead and King (1986). This is a mathematical procedure in which the variability of an M-dimensional object (matrix) is allocated along M orthogonal directions, in decreasing order of variability. Thus, the first singular value represents the largest variation, the second singular value represents the next largest variation, and so on, and the directions of these variation are orthogonal in M-space. The power of this procedure lies in the fact that these directions are not in general the cardinal directions of the space occupied by original object. In other words, given an attractor reconstructed in M dimensions, SVD determines a new set of coordinate axes along which successively decreasing amounts of the attractor exist. The application of this methodology to computations in the state space is perhaps apparent. The SVD parameters can be used to "rotate" the attractor so that it aligns with the newly identified coordinate directions. Those directions along which there is very little variation (those that have small singular values) can be considered to contain only noise, while the attractor dynamics *per se* will exist along the first few directions. Therefore, those coordinate directions with small singular values can be discarded, reducing the dimensionality of the attractor while hopefully retaining its dynamics. Subsequent computations can be carried out in this reduced space, with attendant improvements in computational speed. There will also be a reduction in the overall noise level; since noise is distributed across all dimensions, reducing the embedding dimension will remove those dimensions with more noise and less signal. In addition, we saw above that the statistical variability of dimension estimates increases with embedding dimension, and this also argues for carrying out computations with the smallest legitimate embedding dimension.

Dimension estimation is a tricky business and gone are the days when one could report values of correlation dimension in a cavalier manner as "proof" of chaotic dynamics. All hope is not lost, however, as more powerful techniques have come along to augment and to increase the utility of dimension estimates. Dimension should be seen as one of several parameters that can be used to characterize a dynamic system. Its specific usefulness in estimating the number of state variables that are needed to model a system remains valid. When coupled with surrogate data techniques, the correlation dimension can also address nonlinearity and randomness, along with other hypotheses. It can also, as noted, be very useful in monitoring dimensional changes between different physiological states.

References for Chapter 5

AM Albano, J Muench, C Schwartz, AI Mees, PE Rapp (1988) Singular-value decomposition and the Grassberger-Procaccia algorithm. *Physical Review A* 38:3017-3026.
A Babloyantz, A Destexhe (1986) Low-dimensional chaos in an instance of epilepsy. *Proceedings of the National Academy of Sciences of the USA* 83:3513-3517.
A Babloyantz, A Destexhe (1988) Is the normal heart a periodic oscillator? *Biological Cybernetics* 58:203-211.
JB Bassingthwaighte, LS Liebovitch, BJ West (1994) Fractal Physiology. Bethesda MD: American Physiological Society.
DS Broomhead, GP King (1986) Extracting qualitative dynamics from experimental data. *Physica D* 20:217-236.
M Ding, C Grebogi, E Ott, T Sauer, JA Yorke (1993) Plateau onset for correlation dimension: when does it occur? *Physical Review Letters* 70:3872-3875.
J-P Eckmann, D Ruelle (1992) Fundamental limitations for estimating dimensions and Lyapunov exponents in dynamical systems. *Physica D* 56:185-187.
JD Farmer, E Ott, JA Yorke (1983) The dimension of chaotic attractors. *Physica D* 7:153-180.
P Grassberger, I Procaccia (1983) Measuring the strangeness of strange attractors. *Physica D* 9:189-208.
HS Greenside, A Wolf, J Swift, T Pignataro (1982) Impracticality of a box-counting algorithm for calculating the dimensionality of strange attractors. *Physical Review A* 25:3453–3456.
DT Kaplan, RJ Cohen (1990) Searching for chaos in fibrillation. *Annals of the New York Academy of Sciences* 591:367-74.
D Kaplan, L Glass (1995) Understanding Nonlinear Dynamics. New York: Springer.
B Mandelbrot (1967) How long is the coast of Britain? Statistical self-similarity and fractional dimension. *Science* 156:636-638.

BB Mandelbrot (1983) The Fractal Geometry of Nature. New York: WH Freeman and Co.

F Mitschke, M Möller, W Lange (1988) Measuring filtered chaotic signals. *Physical Review A* 37:4518-4521.

TCA Molteno (1963) Fast O(N) box-counting algorithm for estimating dimensions. *Physical Review E* 48:R3263-R3266.

AR Osborne, A Provenzale (1989) Finite correlation dimension for stochastic systems with power-law spectra. *Physica D* 35:357-381.

E Ott, T Sauer, JA Yorke (eds) (1994) Coping with Chaos: Analysis of Chaotic Data and the Exploitation of Chaotic Systems. New York: Wiley-Interscience.

JB Ramsey, HJ Yuan (1990) The statistical properties of dimension calculations using small data sets. *Nonlinearity* 3:155-176.

PE Rapp (1993) Chaos in the neurosciences: cautionary tales from the frontier. *Biologist* 40:89-94.

PE Rapp, AM Albano, TI Schmah, LA Farwell (1993) Filtered noise can mimic low-dimensional chaotic attractors. *Physical Review E* 47:2289-2297.

PE Rapp, TR Bashore, JM Martinerie, AM Albano, ID Zimmerman, AI Mees (1989) Dynamics of brain electrical activity. *Brain Topography* 2:99-118.

D Ruelle (1990) Deterministic chaos: the science and the fiction. *Proceedings of the Royal Society of London A* 427:241-248.

CE Shannon, W Weaver (1963) The Mathematical Theory of Communication. Champaign: University of Illinois Press.

M Shelhamer (1997) On the correlation dimension of optokinetic nystagmus eye movements: computational parameters, filtering, nonstationarity, and surrogate data. *Biological Cybernetics* 76:237-250.

JE Skinner, CM Pratt, T Vybiral (1993) A reduction in the correlation dimension of heartbeat intervals precedes imminent ventricular fibrillation in human subjects. *American Heart Journal* 125:731-743.

J Theiler (1986) Spurious dimension from correlation algorithms applied to limited time-series data. *Physical Review A* 34:2427-2432.

J Theiler (1990) Estimating fractal dimension. *Journal of the Optical Society of America A* 7:1055-1073.

J Theiler (1990) Statistical precision of dimension estimators. *Physical Review A* 41:3038-3051.

Chapter 6

Surrogate data

We come now to one of the most powerful techniques in our arsenal of computational tools. The methodology is directly applicable to dimension estimation (Chapter 5), nonlinear forecasting (Chapter 7), recurrence analysis (Chapter 8), and indeed as an adjunct to many quantitative measures. It is used to test such hypotheses about a system as the presence of nonlinearity and randomness, and can be extended to test many other specific hypotheses. Although developed for and applied to nonlinear dynamical systems, the concept of surrogate data has potentially very wide applicability. Based conceptually on bootstrap methods in statistics, we introduce the technique of surrogate data by analogy to the statistical t-test.

6.1 The need for surrogates

It should be very apparent from our discussion in Chapter 5 that dimension estimates, although useful, have serious limitations. Among the most serious is the lack of a standard and generally applicable way to assign confidence intervals (error bars) to dimension estimates. This is particularly acute if one is trying to determine if an attractor has non-integer dimension, which is one of the hallmarks of a strange attractor and associated chaotic dynamics.

Surrogate data addresses this problem in an indirect way. Rather than specifying statistical bounds on the dimension estimate from a given signal, bounds are instead established on a *surrogate* signal that has been generated based on some hypothetical process model. The surrogate embodies a hypothesis about the data. By generating many surrogates from a single hypothesis, *empirical* statistical bounds can be placed on the dimensions of the surrogates. Then, the dimension of the original signal can be compared to the distribution of dimensions from the surrogates, and if they overlap, the hypothesis under which the surrogates are generated cannot be ruled out as an explanation for the data. Common hypotheses under which surrogates are generated are that the data are random or are from a linear system. As we shall see, other hypotheses about the underlying system can be tested as well.

Another important role of surrogates is as a numerical control for dimension estimates. Surrogates allow us to ask the question: does some anomalous and unimportant aspect of the signal result in an artifactual value of dimension?

6.2 Statistical hypothesis testing

We approach the concept of surrogate data from the point of view of classical statistical hypothesis testing. The case of statistical testing with the Student t-test is a well-known example. Let us say that we draw a *sample* of items from a larger *population* (the population may be hypothetical, or real but intractably large), and we want to determine if some quantity that is measured on this statistical sample is different from zero. The average (sample mean) of the sample can be found easily, but since it is based on only a subset of the entire population it is a *random variable* that itself has a mean and standard deviation. How far from zero does the sample mean have to be, in order to say that the *true value* of the measured quantity (that for the population as a whole) is not zero?

Statistical hypothesis testing addresses this issue by first assuming a statistical structure for the underlying population, with an attendant null hypothesis. For the t-test, the null hypothesis is that the measured quantity in the population has a normal or Gaussian distribution with a mean of zero (although in general the mean can be non-zero). Now the question is this: given that we have drawn data samples from a larger Gaussian distribution, how likely is it that we would get a sample mean of a given value purely by chance? If this probability is very small, then we can reject the null hypothesis: it is very unlikely that we would get a sample mean this large purely by chance under the null hypothesis that the population is truly zero-mean.

Since we have assumed that the measured quantity in the population has a Gaussian distribution, we can derive analytically the probability distribution of some *test statistic* (the t value), which in this case is the sample mean scaled by the standard deviation. The distribution thus derived is the t distribution, and the test statistic follows this distribution under the null hypothesis. (The reason that the test statistic, based on the sample mean, does not simply follow a Gaussian distribution is because we have drawn samples from a population where the standard deviation is unknown. This impacts the derivation and makes the t distribution slightly broader than the Gaussian.)

Now that we have a value of t, based on the sample mean, we can ask very specifically: how likely is it that we would get our particular value of this quantity, given that it has a t distribution with a hypothesized mean of zero? This can be determined from a table of t distribution values. If the sample mean is large, the test statistic will be large, and it will be unlikely that such a large value could be obtained from sampling a population with a mean of zero. Thus we would reject the null hypothesis.

Note carefully the statistical language. If the observed value is unlikely, based on the null hypothesis, then we can say that our sampled data set is *not consistent with* the null hypothesis: we reject the hypothesis. This *does not* prove that the alternative hypothesis is true. There may be other explanations for the data that we obtained, it is just that we have gone a long way in ruling out one possibility.

This detailed and somewhat pedantic explanation is included here because the surrogate data methodology follows the same line of reasoning.

6.3 Statistical randomization and its implementation

To apply this statistical reasoning to dimension estimates, we create several surrogate data sets, under a single null hypothesis. We find the correlation dimension of each surrogate, and thereby empirically determine the distribution of dimensions under that particular null hypothesis. Then we compare the correlation dimension from the original data to this distribution. If the original dimension is very different from the distribution of dimensions under the null hypothesis, then we can reject the null hypothesis (the hypothesis is not compatible with the dimension of our data). If the original dimension is within the distribution of dimensions from the null hypothesis, then we can not properly reject the null hypothesis.

The key to this usage is that many different surrogates can be generated under a single null hypothesis. This is because surrogates are generated by *randomizing* some aspect of the data, and this randomization can be carried many times with different results each time, but all based on the same null hypothesis.

Perhaps the simplest null hypothesis is that the data are random and drawn from a population with a given set of values. The surrogate in this case (a *shuffle* surrogate) is generated by randomly shuffling the data values. This is done a large number of times to create many surrogates, and the dimension of the original data is then compared to the surrogate dimensions. A slightly more refined surrogate is that the data are random but specifically Gaussian, with a given mean and standard deviation. Surrogates are also commonly generated under the null hypothesis that the data come from a *linear* system with a particular autocorrelation function. These will be presented in turn below.

The surrogates are *random* signals. They are generated by randomizing some aspect of the original signal. Pure noise has infinite dimension. Therefore, the surrogates should have dimensions that are greater than that of the original. In practice, it is common to generate 10 to 100 surrogates of a given type, and to compare the original with the distribution of dimensions from the surrogates. The further the dimension of the original data is from the dimensions of the surrogates, the more justification there is for rejecting the null hypothesis that was used to generate the surrogates. In other words, if the dimension of the original is within the range of dimensions of the surrogates, then we have little basis for saying that the signal was generated by a process that is different from that used to generate the surrogates.

In making this comparison, one might just report how many of the surrogate dimensions are greater than the dimension of the original. Alternatively, we can make use of the probability of different chance arrangements of the resulting dimension values (Theiler & Rapp 1996). Let there be $n-1$ surrogates of a given type, for a total of n dimension values (one from the original signal). There are $n!$ distinct ways to arrange these n dimension values. In $1/n$ of these arrangements, the dimension of the original signal (indeed any pre-selected dimension) will be less than all of the other dimensions. Thus, totally by chance, out of all possible arrangements of the dimensions, the dimension of the original signal will be less than all the others with a probability of $1/n$. If in fact the dimension of the original is less than all of the others, then this could have happened by chance with a probability of $1/n$, and this can be reported as a statistical significance level. As an example, with 39 surrogates, $n=40$, and the significance level is $1/40=0.025$. This represents a one-tailed test, where *a priori* we expect the original dimension to be less than those of the surrogates. If we adjust this to the case when we just want the dimension of the original to be outside the range of dimensions of the surrogates, in either direction, then the significance level is doubled.

While we have described surrogates thus far in terms of their use with the correlation dimension, they can in fact be applied to other measures that we will meet in later chapters, and we will see their use in those applications.

6.4 Random surrogates

If we simply want to test the hypothesis that our signal is a randomly ordered set of fixed values, the *shuffle* surrogate can be used. To generate a *shuffle* surrogate, simply randomly rearrange the values in the signal. The null hypothesis is that the signal is a set of values in no apparent order drawn from a given finite population. This surrogate may seem trivial, but it is a good way to verify that some specific aspect of the distribution of values in the signal is not fooling the dimension algorithm. If the dimensions of these surrogates are not clearly greater than the dimension of the original signal, then we cannot reject the hypothesis that the data can be modeled as a random process with a given set of values. This is the absolute minimum that we should expect of our signal if we expect to investigate it further for dynamical properties.

Slightly more specific is the null hypothesis that the signal is random, with values drawn from a Gaussian distribution with a given mean and variance. The surrogates are generated by drawing values from a Gaussian distribution with the same mean and variance as the original signal. The null hypothesis is equivalent to that of a Gaussian white noise (GWN) process generating the signal.

6.5 Phase-randomization surrogate

Surrogates can be used to test for nonlinearity. Recall from Chapter 3 that the autocorrelation function of a signal describes its linear correlation structure, and that the power spectrum is the Fourier transform of the autocorrelation function. Thus, we can generate a signal that has the *same linear correlations* as the original, but is otherwise random. We do this by retaining the power spectrum of the original signal in the surrogates, while randomizing the phases of the frequency components (the phase spectrum). Since the autocorrelation function carries no information about the phase spectrum, we can randomize phases without affecting the autocorrelation or power spectrum. The resulting surrogates are random, since their phase spectra have been randomized. The null hypothesis is that the signal comes from a linear process with the same autocorrelation function as the original. Rejecting this null hypothesis lends support to the notion that the original signal may come from a nonlinear system.

To generate the phase-randomization surrogate, first take the discrete Fourier transform (DFT) of the signal. Then randomize the phases of the frequency components, either by randomly shuffling them or by replacing them with random phases drawn from (for example) a uniform distribution from 0 to 360 deg. Finally, take the inverse DFT of this frequency-domain signal to obtain a surrogate time-domain signal.

(The phase spectrum of a real (non-complex) time signal has odd symmetry. That is, the phase of the component at frequency f must be the negative of the phase of the component at frequency $-f$. This symmetry should be retained in the randomization.)

This surrogate can give false identifications of determinism (Rapp *et al.* 1994) and should be used in conjunction with other surrogates, as for example the AAFT surrogate introduced next.

6.6 AAFT surrogate

A refinement of the *phase-randomization* surrogate tests the hypothesis that the signal is from a linear system, but then processed by a *static monotonic nonlinear observation function*. That is, there is a nonlinearity in the measurement procedure but not in the system dynamics. This nonlinearity is *static*, meaning that it operates only on the current value of its input – it has no memory or dynamics. It is *monotonic*, which means that while it modifies the amplitudes of the input values, it maintains their rank ordering – the largest of the input values will produce the largest of the output values, and so on. (An example of such a function is logarithmic compression – a log function – applied to a signal from a linear system. The log-transformed data do not enter back into the linear system, which would make the dynamics nonlinear. Instead, the logarithm operation is "tacked on" after the linear dynamical system has generated its signal. A median filter does *not* meet the requirements since it depends on more than just the current value in generating an output, and the absolute value function does *not* meet the requirements since it is not monotonic.)

The AAFT surrogate is formed by first creating a signal from Gaussian random numbers, with the same rank order of the values in time as in the original data. A phase-randomized version of this random signal is formed, as described above. Finally, the original data are rearranged to have the same rank order of the values in time as in the phase-randomized signal just generated. This rearranged data set is the *AAFT* surrogate (amplitude adjusted Fourier transform: Thieler *et al.* 1992.)

The process used to generate this surrogate is non-intuitive, and so we belabor the point to some extent. In slightly more detail, the generation of this surrogate is based on the premise that there is an underlying signal $y(t)$ that is linearly correlated Gaussian noise. This is the hypothesized source of our observed signal $x(i)$ before passing through the nonlinearity. The observed signal $x(t)$ is then $y(t)$ processed by the static nonlinearity $h(\cdot)$: $x(t)=h[y(t)]$. The surrogate $x'(t)$ is obtained by then shuffling the time order of $x(t)$ while preserving the linear correlations of $y(t)=h^{-1}[x(t)]$. The first step rescales $x(t)$ so that the values are Gaussian-distributed. Then the DFT is used to make a signal that has that same autocorrelation function and power spectrum as the rescaled $x(t)$. This Gaussian surrogate is then rescaled to have the same amplitude distribution as the original $x(t)$, generating $x'(t)$.

Finally, we describe this surrogate in excruciating detail, since the logic can be difficult to follow.

1. Generate Gaussian signal (GWN) $y(t)$.
2. Reorder the values of $y(t)$ to match the rank order of $x(t)$. This produces $y_R(t)$, which "follows" $x(t)$ (by effectively applying the *inverse* of the static monotonic transform $h(\cdot)$ that generated $x(t)$, or $h^{-1}(\cdot)$), but is Gaussian. This is meant to represent the hypothesized "original" signal, call it $x_{pre}(t)$, *before it was rescaled by h to become x(t)*.
3. At this point, $y_R(t)=x_{pre}(t)=h^{-1}[x(t)]$.
4. Now take the Fourier transform (typically using the DFT or FFT) of the signal $x_{pre}(t)$ or $y_R(t)$, randomize the phases as above, and take the inverse Fourier transform to get a random time series $y'(t)$. This can be done any number of times to generate as many surrogates as desired. Each $y'(t)$ is a random signal that represents a Gaussian linearly correlated time series before passing through $h(\cdot)$. (Note that $y_R(t)$ and $y'(t)$ are linearly correlated Gaussian signals with identical autocorrelation functions and power spectra but different phase spectra.)
5. Finally, reorder the values of $x(t)$ so that they follow the rank order of the values of $y'(t)$. This produces the surrogate $x'(t)$, which has the same amplitude distribution as the original $x(t)$ but mimics passing a linear Gaussian signal through $h(\cdot)$.

This rather convoluted procedure is necessary because a static nonlinearity will in general change the autocorrelation function and the power spectrum, and therefore multiple surrogates must be generated based on an underlying signal $x_{pre}(t)$ that represents linear Gaussian noise before passing through the nonlinearity $h(\cdot)$. It is helpful to note that the nonlinearity $h(\cdot)$ is not random or arbitrary. The reason that $x(t)$ is reordered to obtain the surrogate $x'(t)$ is so that the function $h(\cdot)$ that was hypothesized to generate $x(t)$ in the first place can be re-applied to the underlying linear Gaussian signals $y_R(t)$ and $y'(t)$, to get $x'(t)$. So, $x'(t)$ is linear Gaussian $y'(t)$ put through $h(\cdot)$. It is the initial re-ordering of $y(t)$ to get $y_R(t)$, by matching the rank order of $x(t)$, that established $h^{-1}(\cdot)$. (See section 19.3 for another approach to understanding this surrogate.)

Although AAFT is a powerful surrogate, problems have been identified in its construction (Stam *et al.* 1998, Schreiber & Schmitz 2000). The most prominent concern involves a signal that is periodic or has strong periodic components. If the length of the signal is such that it does not contain a full integer number of such periods, then a "virtual transient" will be introduced in the time signal, leading to errors in the Fourier transform process when obtaining the power spectrum (see section 3.3). Similarly, if any periodic signal components do not coincide exactly with discrete frequencies in the DFT or FFT, then "smearing" of these components will occur: their energy will appear at several adjacent frequencies. Subsequent randomization of the phases of these frequency components will disrupt the structure of these periodicities. Any phase-randomization surrogate (including the AAFT) of a periodic signal should be another periodic signal, and this will not always be the case if the frequency components are disarranged in this manner. A slightly more complicated iterative surrogate construction scheme has been proposed to remedy these problems (Schreiber & Schmitz 1996). Another approach to surrogates for data with strong periodic components is described in the next section.

6.7 Pseudo-periodic surrogate

The surrogates discussed above are not very useful in assessing signals that have strong cyclic or pseudo-periodic properties. This is because the null hypotheses for these surrogates are obviously false when strong periodicities are present. A more appropriate null hypothesis can be generated, which will allow a more realistic and useful assessment of pseudo-periodic signals. The resulting *pseudo-periodic surrogates* (Small et al. 2001, Small & Tse 2002) test the null hypothesis that the data can be modeled as a periodic process driven by uncorrelated noise (which may be manifest as changes in periodicity from cycle to cycle).

The surrogate to test this hypothesis is generated in the M-dimensional state space. In fact a surrogate *attractor* is generated, which matches the original attractor in the large scale (periodic and near-periodic orbits) but disrupts its small-scale structure. The surrogate-generation algorithm works by first randomly selecting a starting point on the original attractor as the first surrogate point; the next point on the surrogate attractor is the projection one step ahead of a neighbor of the starting point on the original attractor. This is repeated until enough points are generated on the surrogate attractor. In this way the large-scale flow around the attractor is maintained while any correlations between nearby points (noise or local dynamics) are disrupted.

The surrogate is generated by the following procedure:
1. Randomly choose one of the M-dimensional attractor points $x(i)$ as an initial reference point for the M-dimensional surrogate. Call it $s(1)$, the first surrogate point.
2. Set $i=1$.
3. Find all distances from reference point $s(i)$ to the other points $x(\cdot)$.
4. Proportional to their closeness to $s(i)$, randomly select one of these other points $x(\cdot)$, let's say it is $x(j)$.
5. Set the next surrogate point $s(i+1)$ equal to the next point in time after the selected neighbor $x(j)$.
6. Increment i and repeat this process until $i=N$.

7. This creates a surrogate attractor. Now pick off the first component of each M-dimensional surrogate point $s(\cdot)$ as the surrogate time series $y(\cdot)$.

The neighbors are chosen in step 4 with a probability inversely related to their distances from the reference $s(i)$:

$$\text{Prob[choosing neighbor } x(j)] \propto \exp\left[\frac{-|x(j)-s(i)|}{\rho}\right]$$

Selection of the *noise radius* ρ is a key issue. If ρ is too large, then the dynamics of $x(\cdot)$ are not well captured by $s(\cdot)$, because neighbors can be too far from the reference. If ρ is too small, then $x(\cdot)$ and $s(\cdot)$ will be almost identical, since the neighbors will be constrained to be too close to the reference. A recommended choice for ρ is that value that maximizes the number of consecutive point *pairs* (but not longer segments) that are the same in the original data $x(\cdot)$ and the surrogate $s(\cdot)$.

6.8 First differences and surrogates

It is not uncommon to analyze first-differences of a time series $x(i)$ rather than the time series itself: $[x(i)-x(i-1)]$. This procedure is useful in reducing slow drift and other linear correlations that might adversely affect computations (such as correlation dimension) by imposing artifactual temporal correlations unrelated to the shorter-term dynamics that are usually of more direct interest. (This is akin to the temporal dead-zone window presented in Chapter 5 that is intended to reduce temporal correlations in state-space data.) As an example, in investigating the year-to-year dynamics of disease epidemics, it would be helpful to eliminate any long-term trend that might be caused by steady improvements in public health reporting rather than by the disease dynamics themselves (Sugihara & May 1990). Differencing of the data also can increase the density of points in the state space, which may have computational advantages.

For a signal with correlated random components, differencing may increase the dimension since it increases the noise level and removes linear correlations. As discussed in Chapter 5, certain filtered noises can mimic low-dimensional chaos, and differencing reduces or removes these correlations, "whitening" the signal – making it more like white noise which has infinite dimension.

For a deterministic signal such as one that is chaotic, differencing should generally have no effect on the dimension, because any overall long-term dynamics are not altered, only local short-term trends. (Some care must be used because what appears as drift may of course be the consequence of a slow nonlinear process. This takes us back to the discussion of variables and parameters in Chapter 1.) We might think in terms of analogy to the use of position and velocity as state variables for simple harmonic motion as in Chapter 4. If each of these is replaced with its derivative, then velocity and acceleration become the state variables, and these are related to each other in the same way as position and velocity. Thinking of first-differencing as an approximation to differentiation, we can see the soundness of the differencing procedure.

Having established the value of differencing in some cases, we now come to the issue of the appropriate surrogate for differenced data (Prichard 1994). There are two possibilities: generate surrogates based on the original data and then take first-differences of the surrogates, or take first-differences of the data and then generate surrogates based on the differenced data. The general recommendation is that surrogates be generated from the original signal, and then both the original and the surrogates should be differenced. Comparisons should then be made between the dimensions (or other derived quantities) of these differenced signals. This is especially pertinent to the AAFT surrogate which involves a static nonlinear transformation, since differencing and the nonlinear transformation are not necessarily commutative (the order in which they are applied matters, unlike the case for linear operations).

6.9 Multivariate surrogates

In section 4.7 we discussed the fact that inherently multivariate signals can be used in investigations of nonlinear dynamics. Examples include EEG or EKG signals recorded with multiple electrodes. In these cases it is useful to have surrogates that incorporate hypothesized relations between signals. Such multivariate surrogates can be generated (Prichard & Theiler 1994).

These surrogates are based on the hypotheses that each signal can be modeled as linearly correlated Gaussian noise, and that there are only linear correlations between the signals. These inter-signal correlations can be described by cross-correlation functions, just as linear correlations within a signal are described by the autocorrelation function.

The cross-spectrum $S_{xy}(f)$ of two signals $x(t)$ and $y(t)$ is the Fourier transform of their cross-correlation function. It can be shown (although we will not do so) that if the power spectrum and phase spectrum of a signal $x_i(t)$ are, respectively, $A_i(f)$ and $\phi_i(f)$, then the magnitude (power) and phase components of the cross-spectrum of the signals $x_i(t)$ and $x_j(t)$ are:

$$|S_{x_i x_j}(f)| = A_i(f) A_j(f)$$
$$\angle S_{x_i x_j}(f) = \phi_j(f) - \phi_i(f).$$

To retain the linear correlations *within* a signal $x_i(t)$ as well as the linear correlations *between* signals $x_i(t)$ and $x_j(t)$, it is necessary to retain the cross-spectrum for all pairs of signals (for all i,j pairs). At the same time we need to randomize the autocorrelation function of each individual signal $x_i(t)$. We can do both of these by adding the *same* random set of phases to the phase spectrum of each of the $x_i(t)$. In other words, for each signal $x_i(t)$, generate a phase-randomization surrogate as described in section 6.5. Randomize the phases in each case by adding a random set of phases to the phases obtained by taking the Fourier transform of each signal, adding the same random set of phases to all of the phase spectra. Since the linear correlations between signals are embodied in the cross-correlation function, and the cross-spectrum determines the cross-correlation, then we can retain the cross-correlations by making sure that all cross-spectra are unchanged. This is achieved by generating surrogate signal pairs such that the phase portion of the cross-spectrum of each signal pair is not altered, and this is easily accomplished by modifying each individual phase spectrum $\phi_i(f)$ and $\phi_j(f)$ by the same amount so that their difference $\phi_j(f) - \phi_i(f)$ is not changed. This is the result of the surrogate generation process described here.

6.10 Surrogates tailored to specific physiological hypotheses

By far the most frequent application of surrogates is to test for nonlinearity and randomness, and the surrogates described above are in common use for this application. One should not rule out the possibility, however, of generating other surrogates to address specific hypotheses about a given physiological system.

As an example, Theiler (1995) studied surrogates generated from epileptic EEG waveforms. The signal consists of near-periodic spikes, with a brief wave-like pattern between each spike. Typically, the dynamics of spike timing are assessed by generating various surrogates based on inter-spike timing, ignoring the morphologic details of the signal. In this study, on the other hand, any dynamics embodied in the shape of the waveform itself were also of interest. To test the null hypothesis that there is no dynamical correlation from one spike-plus-wave to the next, surrogates were created by dividing the signal into segments, each segment containing one spike and the subsequent wave, and these segments were scrambled.

Kaplan and Glass (1993) took a similar approach to the generation of surrogate data on measles cases, by creating annual peaks with random amplitudes to represent the number of reported measles cases.

Another example (Shelhamer 1997), which we will explore in more detail in Chapter 15, is the analysis of the structure of optokinetic nystagmus eye movements (OKN, described in Chapter 4). From Fig. 4.5.1A we can see that OKN is composed of alternating slow and fast phases. If we want to test the hypothesis that the properties of a fast phase are determined by the properties of the preceding slow phase, for example, then we can create a surrogate by separating the OKN signal into segments, each containing a fast phase and the preceding slow phase, and these segments can be put back together in random orders to generate surrogates. Then the dimensions of the surrogates can be compared to that of the original OKN. If the surrogate dimensions are greater, then randomness has been introduced by this scrambling process, and it is not solely these slow-fast segments that determine the dimension.

6.11 Examples of different surrogates

Examples of four different types of surrogates were generated from the Lorenz x variable and are shown in Fig. 6.11.1. Clearly the pseudo-periodic surrogate most closely resembles the original signal, because of its retention of cyclic structure. The phase-randomization surrogate is next closest qualitatively to the original, because it retains linear correlations in time.

The correlation dimension was determined for each surrogate type. Normally, several surrogates of each type would be generated and their dimensions compared to that of the original signal. In this case, however, we present dimension results from only a single example of each surrogate type; the results are typical of those from multiple realizations of each surrogate. The results of the correlation dimension computations are shown in Fig. 6.11.2, for the surrogates given in Fig. 6.11.1. The results can be compared to those from the original Lorenz data in Fig. 5.4.1.

For the *shuffle* surrogate (top row), the slope of the correlation integrals increases as embedding dimension M increases from 5 to 10. The dimension computation does not converge, and no dimension can be assigned to this signal. (The computations could be extended to larger values of M, but even at $M=10$ any possible scaling region is very small (flat portion of top curve in the right column), and this trend does not improve as M increases.) This is consistent with a random signal, as we would expect from this surrogate.

The same result is obtained from the *Gaussian* surrogate (second row), again as expected.

The *phase-randomization* surrogate (third row) produces a different result. First, any possible scaling region is quite small, less than 0.25 $\log_{10}$ units. Second, to the extent that the slope of the correlation integrals converges, the asymptotic value is approximately 5.5, which is much greater than the correlation dimension of the Lorenz x variable itself, as we know from Chapter 5. Thus we have confirmed that the Lorenz system cannot be described as linearly correlated noise; linear correlations within the data are not sufficient to reproduce the dimension and hence a linear model of the process is not satisfactory.

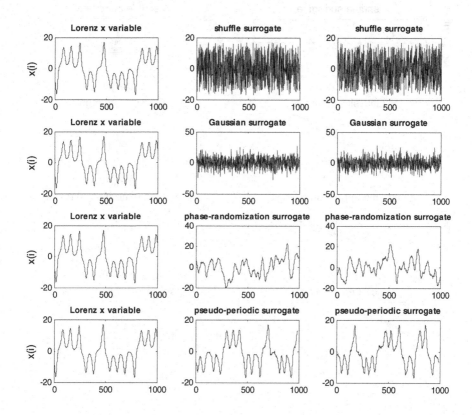

Figure 6.11.1. Examples of surrogate data generated from 1000 values of the x variable of the chaotic Lorenz system (left column). In each row are two examples of four different types of surrogate data as described in the text. The *shuffle* surrogate is generated by shuffling the data values, the *Gaussian* surrogate consists of values from a normal distribution with the mean and standard of the original signal (note change of scale), the *phase-randomization* surrogate is generated by randomizing the phases of the frequency spectrum thereby preserving linear correlations in time, the *pseudo-periodic* surrogate maintains cyclic structure and its generation is described in the text.

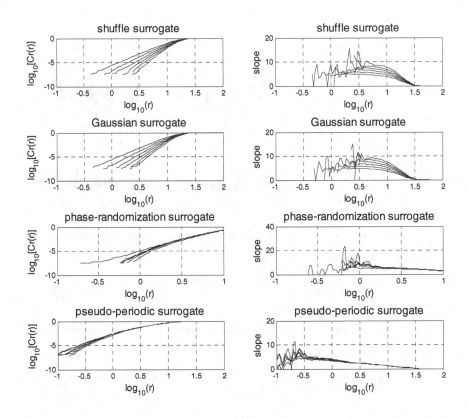

Figure 6.11.2. Examples of correlation dimension computations carried out on surrogate data sets generated from the Lorenz system. Each row shows results for one sample of one of the surrogates shown in Fig. 6.11.1. Correlation integrals are shown on the left, and slopes on the right, for embedding dimensions from 5 to 10.

Finally, the *pseudo-periodic* surrogate (bottom row) comes closest to the results from the original Lorenz data. However, the scaling region for the surrogate is much smaller, and the mean slope (D_{corr}) in the scaling region is 2.31, much greater than that of the original data. This confirms that the Lorenz system is not periodic with uncorrelated noise disrupting the periodicity. Rather, the variations in period from cycle to cycle are inherent in the deterministic dynamics.

6.12 Physiological examples

Certainly the earliest systematic treatment of the concepts of surrogate data is the one cited above (Theiler *et al.* 1992). Shortly thereafter the first physiological applications appeared. Among them is a brief report on the correlation dimension of OKN (Shelhamer 1992), where an "artificial OKN" signal was examined as a type of surrogate data. This surrogate was created by generating an idealized sawtooth waveform, with fast phases randomly inserted; the mean and variance of the fast-phase amplitudes and intervals were identical to those of the original OKN. The dimension computations for this surrogate did not converge, while OKN itself had a dimension of about 2.8.

An earlier (possibly the first) and more extensive physiological application of surrogates was to the H-reflex (Schiff & Chang 1992). This is a monosynaptic spinal cord reflex, electrically stimulated at the tibial nerve, with muscle activity measured at the gastrocnemius with surface EMG. Stimulus pulses were applied once per second, and five sequences of 1024 responses were recorded. The data series consists of the peak amplitude of consecutive responses. Correlation dimensions were computed for the original data sets and for surrogates based on them. The surrogates were the three types described above: shuffle, Gaussian, phase-randomization. The study was designed to look for dynamical correlations between the amplitudes of consecutive responses.

Unfortunately the raw data series did not converge to a dimension value, so a 21-point moving average (MA) was applied (MA is a simple form of low-pass filtering, where each data point is replaced with the average of the surrounding points). The MA procedure was also carried out on the three types of surrogates. The dimension computations for the shuffle and Gaussian surrogates did not converge – the values increased with embedding dimension up to $M=11$. The MA experimental data converged to a dimension of 2.6 ± 0.02, while the MA phase-randomized data yielded a value of 2.3 ± 0.004. (This is an unusual case in which the surrogate has a lower dimension than the original data.) The conclusion is that the sequence of H-reflex amplitudes cannot be distinguished from linearly correlated noise; there is no apparent nonlinear dynamical relationship of one response to the next.

References for Chapter 6

DT Kaplan, RJ Cohen (1990) Searching for chaos in fibrillation. *Annals of the New York Academy of Sciences* 591:367-74.

DT Kaplan, L Glass (1993) Coarse-grained embeddings of time series: random walks, Gaussian random processes, and deterministic chaos. *Physica D* 64:431-454.

E Ott, T Sauer, JA Yorke (eds) (1994) Coping with Chaos: Analysis of Chaotic Data and the Exploitation of Chaotic Systems. New York: Wiley-Interscience.

D Prichard (1994) The correlation dimension of differenced data. *Physics Letters A* 191:245-250.

D Prichard, J Theiler (1994) Generating surrogate data for time series with several simultaneously measured variables. *Physical Review Letters* 73:951-954.

PE Rapp, AM Albano, ID Zimmerman, MA Jiménez-Montaño (1994) Phase-randomized surrogates can produce spurious identifications of non-random structure. *Physics Letters A* 192:27-33.

SJ Schiff, T Chang (1992) Differentiation of linearly correlated noise from chaos in a biologic system using surrogate data. *Biological Cybernetics* 67:387-393.

T Schreiber, A Schmitz (1996) Improved surrogate data for nonlinearity tests. *Physical Review Letters* 77:635-638.

T Schreiber, A Schmitz (2000) Surrogate time series. *Physica D* 142:346-382.

M Shelhamer (1992) Correlation dimension of optokinetic nystagmus as evidence of chaos in the oculomotor system. *IEEE Transactions on Biomedical Engineering* 39:1319-1321.

M Shelhamer (1997) On the correlation dimension of optokinetic nystagmus eye movements: computational parameters, filtering, nonstationarity, and surrogate data. *Biological Cybernetics* 76:237-250.

M Small, D Yu, RG Harrison (2001) Surrogate test for pseudoperiodic time series data. *Physical Review Letters* 87:188101-1:4.

M Small, CK Tse (2002) Applying the method of surrogate data to cyclic time series. *Physica D* 164:187-201.

CJ Stam, JPM Pijn, WS Pritchard (1998) Reliable detection of nonlinearity in experimental time series with strong periodic components. *Physica D* 112:361-380.

G Sugihara, RM May (1990) Nonlinear forecasting as a way of distinguishing chaos from measurement error in time series. *Nature* 344:734-741.

J Theiler (1995) On the evidence for low-dimensional chaos in an epileptic electroencephalogram. *Physics Letters A* 196:335-341.

J Theiler, S Eubank, A Longtin, B Galdrikian, JD Farmer (1992) Testing for nonlinearity in time series: the method of surrogate data. *Physica D* 58:77-94.

J Theiler, PE Rapp (1996) Re-examination of the evidence for low-dimensional, nonlinear structure in the human electroencephalogram. *Electroencephalography and Clinical Neurophysiology* 98:213-222.

Chapter 7

Nonlinear forecasting

Soon after the shortcomings of the correlation dimension started to become apparent, other approaches to the characterization of nonlinear dynamical systems began to be developed. Among them, *nonlinear forecasting* or *prediction* is among the most widely used. It is based on a simple principle: a chaotic system is *deterministic*, and therefore it should exhibit some *predictability* in the short term, even though this should decay rapidly. In fact this is one of the defining features of chaos: impossibility of long-term prediction. The presence of short-term predictability can be understood by reference to some of the state-space representations that we have seen so far; trajectories tend to be aligned in many regions of the state space, and so knowledge of the flow of one path might provide information on the flow of adjoining paths. This is the essence of the technique. Different types of systems exhibit different "signatures" of predictability, and this can be used to distinguish between different behaviors. Surrogate data techniques can be applied to nonlinear forecasting as well.

Unlike the correlation dimension, forecasting results are often readily interpretable and point directly to the deterministic structure of the system under study. Useful results can often be obtained with much smaller data sets than is the case with dimension estimates (hundreds rather than thousands of points).

7.1 Predictability of prototypical systems

Details of the forecasting method will be given in the following section. In brief, a reference point is selected on the attractor in state space. A set of trajectory paths close to that point is found, and these neighboring paths are used to create a local linear approximation to the trajectory flow. This approximation is then used to forecast the path of the reference point at future time steps. As the forecast is carried out further into the future, its quality will decay: the forecasts will become less accurate as they move further away from the reference point in time. The nature of this decay of forecasting quality is dictated by the dynamics of the system.

Some examples are shown in Fig. 7.1.1. A purely random system (noise) has uniformly poor forecasting for all time steps. A "conventional" deterministic system, such as a sine wave, has uniformly good forecasting fairly far into the future. Adding noise to such a deterministic system degrades the quality of the forecasting equally across all time steps: the forecasting-quality curve is shifted downward. A chaotic system (the logistic equation from Chapter 1, in the chaotic regime) has good short-term forecasting that decays rapidly. Distinguishing between these broad classes of behavior on the basis of the forecasting signature is fairly straightforward. (There are complications, however, as certain types of noise also exhibit good short-forecasting that decays rapidly; see section 7.6.)

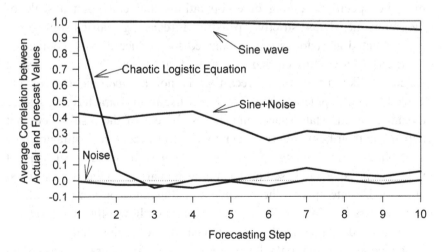

Figure 7.1.1. Forecasting results from different systems. Forecasting quality is shown as a function of the number of steps into the future at which the forecast is performed.

The forecasting procedure can be applied to relatively small data sets, which has great advantages for the analysis of physiological systems, which can be significantly nonstationary. As with the correlation dimension, there are cases in which *relative* rather than *absolute* measures of forecasting quality are valuable: indeed there are cases in which it may be more fruitful to ask "how predictable" a system is, rather than if it is random or deterministic *per se* (Rapp 1993).

7.2 Methodology

The forecasting process is carried out on an attractor that has been reconstructed with the time-delay procedure from Chapter 4. (A theoretical note. The Takens embedding theorem shows that *topological* properties of an attractor are preserved through time-delay reconstruction. Topological properties are based on concepts such as the neighborhood of a point and connectedness of points. A *metric* or measurement scale is not prescribed. Indeed topological transformations can involve stretching and bending operations, which clearly would change distances between points and thus affect any measurement scale. In order to carry out the forecasting procedures, distances between points must be specified, among other operations that call upon a scale of measurement. Thus strictly speaking forecasting carried out on a reconstructed attractor is not guaranteed to produce results identical to those from forecasting carried out on the true attractor in the actual state space. Unlike dimension, forecasting is not a *topological invariant*. Nevertheless, in practice the procedure is useful in characterizing system dynamics in the state space, and the usefulness of the results does not depend on their being referred back to a hypothetical "true" attractor.)

Nonlinear forecasting (prediction) is a way to forecast subsequent values in a sequence, based on local approximations to the reconstructed attractor in state space. The procedure is illustrated on a portion of the Lorenz attractor, in Fig. 7.2.1. The top graph shows 400 points of the reconstructed attractor; the box indicates the region that has been expanded and plotted in the lower graphs. In the lower graphs, individual points are indicated by small circles, and the general directional flow of the trajectory paths is indicated by the arrow to the lower right. A *reference point* has been denoted with a bold **X**; forecasting is carried out from this starting point, and the forecasts estimate the future path of this point. Those points on nearby paths of the attractor that are closest to the reference point have been identified and marked with large bold circles. In this case three such *nearest neighbors* have been identified. Note that these nearest neighbors are not selected from the same trajectory path on which the reference point lies; this would place them subsequent in time to the reference point, making the forecast a temporal extrapolation, rather than a dynamical approximation as desired.

Each of the nearest neighbors is followed one time step into the future; these new points are indicated also with bold circles in the lower right graph. The nearest neighbors and their time course provide a template or model on which to make a forecast of the future time course of the reference point itself. A *linear regression* is carried out on the set of nearest neighbors and their one-step-ahead projections; this creates a *local linear approximation* to the attractor. The resulting linear regression line is shown as a thick line with thick filled circles in the lower right graph; the regression has been carried out for three time steps into the future, starting from the initial reference point.

The one-step ahead forecast in this case is excellent: the forecast point on the bold regression line lies directly on top of the corresponding point on the attractor (the projection one time step ahead of the reference point). After that, since the forecasts follow the regression line in a straight path while the attractor curves upward, the forecasts become less accurate (the regression line deviates from the attractor). Forecasts two, three, and more steps into the future become progressively worse.

This process is repeated for a large number of reference points on the attractor, each time selecting new nearest neighbors and creating new linear regressions. The larger the number of reference points, the better the statistics of the forecasting quality.

To quantify the quality of the forecasts, a comparison is made between the actual values extending from each reference point, and the corresponding forecast values based on that point. This is accomplished by taking, for all forecasts carried out one step ahead, over all reference points, the set of *actual* one-step ahead values and the set of one-step ahead *forecasts*, and finding the correlation coefficient r between these two sets. This is then carried out for two-step ahead values and forecasts, and so on. The behavior of the correlation coefficient as forecasting is carried further into the future, away from the reference point, indicates the nature of the system under study.

A key parameter that must be chosen, in addition to those involved in the time-delay reconstruction itself (delay L and embedding dimension M), is the number of nearest neighbors. Too large a set of neighbors risks using neighbors that are not actually *near* the reference point, and so do not reflect the local flow of the attractor. Too small a set risks not having enough information with which to generate an adequate regression line. While different guidelines exist, choosing a number of nearest neighbors approximately equal to the embedding dimension M is a reasonable starting point. As always, any *a priori* knowledge of the system should guide this choice, and a compulsive investigator might try several values and show that the results are not highly dependent on the specific value.

It is helpful to keep in mind, in selecting parameter values and making other choices regarding the forecasting method, that we are *not* usually interested in optimizing the forecasting *per se*. Rather, we are most often interested in showing the decay in forecasting quality with time, and comparing forecasting ability from different systems or different situations. Thus an exhaustive search for the optimal forecasting algorithm is usually unnecessary.

It is common and highly desirable in forecasting to choose nearest neighbors from one set of attractor points, and reference points from a different set of points. This is typically accomplished by dividing the attractor into two segments in time, an early segment and a late segment. The early segment is treated as a *template* of points (and trajectory paths) – it is this template on which the forecasts are based. The reference points, from which the forecasts emanate, are selected from the late segment. This ensures that forecasts are made *out-of-sample* rather than *within-sample*; the latter could artificially bias the quality of the forecasts. One approach to testing for stationarity of the data set is to use the early segment as a template for forecasts on the later segment, and vice versa, and then compare the results. If the system is stationary, then the two sets of forecasts should be approximately identical.

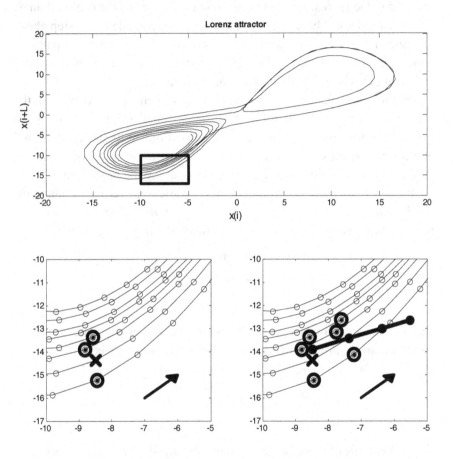

Figure 7.2.1. Representation of the forecasting procedure carried out on a portion of the Lorenz attractor (boxed area in top graph, enlarged in bottom graphs). Scales on all graphs are identical. See text for description of the forecasting procedure.

The forecasting procedure described above makes use of a single regression line, and subsequent points on this line form the forecast values at subsequent time steps into the future. Another way to generate forecasts is in an incremental manner. In this case, the initial forecast is carried out only one time step into the future. Then, this one-step ahead forecast value is treated as an actual point on the attractor (rather than as a forecast value); it becomes the reference point for a new one-step ahead forecast, and so on. This generates a piecewise linear trajectory path, on which the forecast points lie. This process is computationally more involved but will generally produce better forecasts.

In the example above the embedding dimension is two, and the regression line is the familiar two-dimensional version $y=mx+b$. In general the forecasting procedures are carried out in higher-dimensional embedding spaces. The generalization of linear regression to M dimensions is straightforward, and we now outline the procedure.

It is perhaps best to formulate the problem in terms of matrices. Again using the nomenclature of Albano *et al.* (1988), consider the points on the reconstructed attractor as rows of a *trajectory matrix*:

$$A = \begin{bmatrix} \langle - & y_1 & - \rangle \\ \langle - & y_2 & - \rangle \\ & \vdots & \end{bmatrix}_{N \times M}$$

The reference point is one of the points y_i, call it y_{ref}. The k nearest neighbors of y_{ref} are identified and placed into a new matrix:

$$P = \begin{bmatrix} \langle - & y_{neighbor1} & - \rangle \\ \langle - & y_{neighbor2} & - \rangle \\ & \vdots & \end{bmatrix}_{k \times M}$$

The projections of these neighbors one step ahead gives a third matrix:

$$Q = \begin{bmatrix} \langle - & y_{projection1} & - \rangle \\ \langle - & y_{projection2} & - \rangle \\ & \vdots & \end{bmatrix}_{k \times M}$$

We desire to find a set of coefficients that, when applied to the nearest neighbors in P, will produce (with minimum squared error) the projected points in Q. That is, we want to solve this equation:

$$PC = Q$$

where C is the MxM matrix of coefficients. Since we can choose the number of nearest neighbors arbitrarily, this represents a set of under-determined or over-determined equations, which must be solved with least-squares techniques. While we will not go into detail on this point, there are many software packages that will solve this problem. (In MATLAB, for example, the command **C=P\Q** will provide a solution, the backslash operator being shorthand for the least-squares procedure.)

Finally, we apply the set of coefficients to the reference point to obtain the one-step ahead forecast:

$$y_{forecast} = y_{ref} C.$$

7.3 Variations

There are many ways in which forecasting can be carried out, although the one just described appears to be among the most common, at least in its essentials. As noted, one may generate a single linear regression for each reference point, or a set of regressions each one based on the previous forecast value. It is also possible to use more sophisticated local approximations instead of linear regression, such as local polynomials or neural networks.

In the description above, the forecast was based on a set of nearest neighbors to each reference point. A more stringent requirement is to select the smallest *simplex* for a given reference point (Sugihara & May 1990). This is a "ball" of k points with the smallest diameter which *contains* the reference point. That is, the reference point is surrounded by the points that form the simplex. In the simpler case of nearest neighbors, all the neighbors could in general be on one side of the reference point, for example, which might not provide as good a model for the path of the reference as would a simplex. Having identified a simplex or other set of neighbors, we might also weight their contributions to the forecast by their distance from the reference point; neighbors that are closer to the reference point would contribute more to the forecast than neighbors farther away. With the simplex method, the forecast point is determined as the "center of mass" of the set of points in the simplex, projected one step ahead.

Finally, we have used the correlation coefficient r to quantify forecasting quality, but a measure of forecasting error can be used instead, such as mean-squared error.

The methods and variations presented here are adequate for many investigations. Much more general and sophisticated techniques are available (Weigend & Gershenfeld 1993, Casdagli 1989).

Forecasting quality has also been used to determine embedding dimension M, in a manner similar to the use of the False Nearest Neighbors method in Chapter 4 (Sugihara & May 1990). Forecasting quality (r) at one step into the future is determined as a function of M. That is, forecasting is carried out on the same system, with the attractor reconstructed in increasingly higher embedding dimensions. When forecasting quality reaches a plateau or a peak, this indicates that an appropriate embedding dimension has been reached.

7.4 Surrogates, global linear forecasting

Since linear systems produce attractors that can be analyzed with the methods here, the demonstration of forecasting ability does not allow a claim to be made for nonlinearity in the system. (Note that a local linear approximation to an attractor does not imply that the system is linear. Even a simple periodic oscillator has a curved trajectory in state space, which would not be well-modeled by a local linear fit.)

To this end, surrogate data techniques from Chapter 6 can be usefully applied to forecasting. If forecasting quality is unchanged in a given type of surrogate, then the hypothesis behind that surrogate cannot be rejected as an explanation for the data. Often it is sufficient in these comparisons to show forecasting plots (r versus prediction step) for the original data and the surrogates; if the difference is large enough it may be apparent without recourse to statistics. At other times, however, it may be necessary to perform statistical analyses on the surrogates, and in this case a sample of one-step forecasting quality ($r(1)$) can provide a useful quantity for statistical testing.

Another way to approach the question of nonlinearity is to compare nonlinear forecasting as presented here with results from a linear estimation method. One such method is to fit an ARMA model (autoregressive, moving-average, also known as Box-Jenkins models: Box *et al.* 1970). This models each data value as a linear combination of past data values (autoregressive) and values of a hypothetical (typically Gaussian white noise) input process (moving average):

$x(i) = a_1 x(i-1) + a_2 x(i-2) + a_3 x(i-3) \ldots + b_0 u(i) + b_1 u(i-1) + b_2 u(i-2) \ldots + n(i)$.

Here $n(i)$ is a noise term. Since all of the data are used in determining the model parameters a_i and b_i, the procedure is a *global* rather than a *local* fit. Thus it is possible to obtain good-quality forecasts even in nonlinear systems because all data are used, as opposed to local approximations. Keeping the order of the model (the number of terms) low in comparison to the number of data points helps in making a fair comparison with nonlinear forecasting, but in general good-quality forecasting from an ARMA model, with low residual error, indicates a high degree of linear structure in the data. There are established criteria for making this determination (Akaike 1979).

7.5 Time-reversal and amplitude-reversal for detection of nonlinearity

A type of surrogate data for testing of nonlinearity is based on reversibility of the data (Stam *et al.* 1998, Diks *et al.* 1995). The procedure was designed to address some problems associated with the AAFT surrogate (see end of section 6.6). These methods were developed for the case of periodic data (*limit cycles* in the case of nonlinear systems: essentially, periodic attractors), including however such cases as the Lorenz system which is nearly periodic over a limited range. The important point is that the probability distributions of the data values be approximately symmetric in amplitude (about the mean) and in time (moving forward and backward from a given point). Obviously periodic signals meet these criteria most strongly.

The method is referred to as *nonlinear cross-prediction*. A set of *amplitude-reversed* data is generated from the original signal, by removing the mean, inverting the data, and adding the mean back. A set of *time-reversed* data is also generated, by reversing the time order of the data values. Nonlinear forecasting is performed on the original data. Forecasting is quantified by correlation coefficient, comparing forecast and actual data (as usual). In addition, *the same forecast values* are compared to the amplitude-reversed data and to the time-reversed data. If the forecasting quality (correlation coefficient) of the amplitude-reversed data is degraded then the data are not amplitude-reversible, and if the forecasting quality of the time-reversed data is degraded the data are not time-reversible. Lack of amplitude symmetry about the mean leads to poor forecasting of amplitude-reversed data, and lack of time reversibility leads to poor forecasting of time-reversed data.

A static nonlinear transformation will not affect time-reversal properties of the data (since the data will be approximately identically distributed before and after any selected point in time, and the static transformation will affect these two distributions identically). Linearly correlated Gaussian noise is time-reversible in the same sense. Therefore, if the data are not time-reversible, then we can rule out the possibility that the data were created by a static transformation of a linear Gaussian process. On the other hand, a static nonlinear transformation (in general) destroys amplitude symmetry, leading to lack of forecasting in the amplitude-reversed data, while forecasting of the time-reversed data remains good.

A true *dynamic nonlinearity* leads to time irreversibility, and poor forecasting of time-reversed data (with or without a change in forecasting of amplitude-reversed data). However, not all nonlinear systems are time-irreversible; strong asymmetry is an indicator of nonlinearity, but lack of symmetry does not imply linearity.

7.6 Chaos versus colored noise

We have seen in Chapter 5 that certain types of filtered noise, or colored (as opposed to white) noise, can "fool" the dimension algorithms. That is, a finite and possibly non-integer dimension can be produced from random data with certain temporal correlations. Such a problem exists with nonlinear forecasting as well.

The specific types of noise that concern us here are those with a *power-law spectrum*. That is, the power spectrum decays in a manner proportional to frequency raised to some power: $S_{xx}(f) \sim 1/f^{\alpha}$, where $1<\alpha<3$; α can be non-integer. The corresponding autocorrelation function also decays as a power law, rather than as an exponential which is more common: $R_{xx}(\tau) \sim 1/\tau^{\beta}$. These noise processes are known as *random fractal sequences*, or *fractional Brownian motion* (fBm). They are fractals (see Chapter 5) in the sense that they are *self-similar*, but in a statistical sense: statistical properties such as variance or range are related over different time scales by way of a power law. The properties of fBm are interesting in their own right, and we will see some of them in Chapter 13.

For our present purposes, we are concerned with distinguishing between chaos and fBm. This is an important issue because, like chaos, fBm can produce forecasting that is good in the short term but decays rapidly as forecasting is carried further into the future (see Fig. 7.1.1).

Correlations between points decay as a power law for fBm, and this is reflected also in forecasting, where the decay of forecasting quality decays as a power law. A power law expressed on a log-log graph produces a straight line, and so $\log(1-r(T))$ versus $\log(T)$ is a straight line for fBm (T is the number of time steps into the future over which the forecasting is carried out, and $r(T)$ is the average correlation coefficient at each forecasting time step T). On the other hand, a chaotic system has exponential decay of correlations between points, due to exponential divergence of trajectory paths, and this is reflected as an exponential decay in forecasting quality: $\log(1-r(T))$ versus T is linear. Thus the decay in forecasting quality is different for these two processes. These results were developed by Tsonis and Elsner (1992), based on work by Wales (1991). Such a diagnostic criterion has proven to be quite useful (e.g., Shelhamer 2005).

Another approach to this problem has been proposed (Ikeguchi & Aihara 1997). Forecasting is carried out just one step ahead. The conventional correlation coefficient r is used to assess how well the forecast data match the actual data. In addition, a *difference correlation coefficient* is computed, which compares the first differences of the same forecast and actual data. Since forecast data tend to follow trends in the original data, and first differences of power-law noise show more point-to-point variability, the conventional correlation will be high for both signals while the difference correlation will be generally much lower (and even negative) for the noise signals.

7.7 Forecasting of neural spike trains and other discrete events

Forecasting of neural spike trains (and other point processes where the values represent times of occurrence of stereotyped events) presents some special problems. The data values are not evenly spaced in time, and forecasting carried out directly on inter-spike intervals (ISIs) actually shows the behavior of forecasting quality as a function of event number rather than time *per se*. This can complicate the discrimination of chaos versus colored noise, for example, as described above.

One approach to this problem is to generate a smooth function that represents instantaneous event rate, which can then be sampled evenly in time. Such a function is described in section 7.8.

Another issue arises in attempting to use nonlinear forecasting techniques to determine the dynamics of a signal (such as cell membrane potential) that drives the event generator. In Chapter 4 we described how neural inter-spike interval (ISI) data can be used to reconstruct an attractor with the time-delay method. The same study that demonstrated this (Sauer 1994) also showed that when a simple integrate-to-fire neural model is driven by a random signal or by chaos, the resulting ISI data have different forecasting qualities. In particular, the rate of decay of forecasting quality is able to determine whether the input signal to the neuron model is random or chaotic, as described previously in this chapter. An underlying chaotic signal generates chaotic ISIs, and a random signal generates random ISIs, and these can be distinguished by their forecasting signatures.

However, and most crucially, this forecasting ability decreases as the "neuron" firing threshold increases – that is, as the spikes become further apart in time. This makes sense, since the spikes may be less evenly spaced in time as their average rate decreases, and in any case they reflect the underlying dynamics of the integrated signal more poorly as their average rate decreases (in analogy to the Shannon sampling theorem). Intuitively, the intervals should be shorter than the time scale determined by the "forecasting horizon" (time into the future over which reasonably good forecasting can occur) of the underlying chaotic signal. Thus chaotic dynamics in an underlying signal are reproduced in the spike train if average firing rate is not too low, and if firing occurs over the entire range of the input signal so that there is no dead zone (Racicot & Longtin 1997). There appears to be no general solution to the problem of how random or chaotic inputs might be reflected in spike trains if the spike generator does not follow these rules.

7.8 Examples

Instructive and insightful examples of nonlinear forecasting applied to disease epidemics can be found in Chapter 20. Here we discuss a set of studies on neural dynamics.

Dopamine neurons in the substantia nigra often exhibit an irregular firing pattern. They fire more regularly *in vitro*, but the irregular activity can be reproduced in a bath of apamin, which blocks calcium-activated potassium channels. Nonlinear forecasting was carried out on these spike trains, to investigate the dynamics of the neural firing. To avoid the problems noted above with unevenly spaced data values, a spike-density function (SDF) was created from each data set by placing a Gaussian (bell-shaped) curve at the time of each spike, and adding these functions. Since they overlap when firing rate is high, this creates a smooth function that represents instantaneous spike density or firing rate. This smooth function is then sampled to produce a time series with evenly spaced data points. Forecasting of the SDF was carried out using procedures described by Sugihara and May (1990) and referred to previously, including the construction of a simplex of points for each reference point, and generating the forecast values through a weighted average of the simplex points as they move forward in time. An embedding dimension of three was used, and data sets contained several hundred spikes.

The initial study of these data (Lovejoy *et al.* 2001) included careful and extensive simulations with artificial periodic data, chaotic data, and data from a neural model in a near-periodic pacemaker mode and in a chaotic mode. A Poisson-distributed random variable was used as noise. Forecasting quality for pure Poisson noise decayed the most rapidly, with some forecasting ability due to inter-point correlations imposed by the SDF. Driving the neural model in the pacemaker mode with Poisson noise resulted in forecasting that decayed approximately linearly (r vs. forecasting time T), consistent with correlated noise. The model driven with Poisson noise but in the chaotic regime yielded exponential decay ($\log(1-r)$ vs. T is linear), consistent with the chaotic nature of the model.

With actual neural data, irregular firing *in vivo* and apamin-induced firing *in vitro* both showed an exponential decay of forecasting quality r, consistent with chaos. An *in vitro* control (no apamin) showed linear decay of r, as for noise. From this the authors concluded that there is deterministic chaotic firing *in vivo*, and with apamin *in vitro*, and that this pattern of activity is not necessarily the result of random synaptic inputs but rather stems from dynamics intrinsic to the neuron.

However, a later study by two of the same authors revisited this result (Canavier *et al.* 2004). In this later study, surrogate data sets were generated in an attempt to confirm the earlier finding of chaotic activity. However, shuffle surrogates (generated by randomizing the order of the spike intervals) showed the *same* exponential decay of forecasting quality r as did the actual neural data, for both the apamin *in vitro* and the spontaneous *in vivo* irregular firing. On the other hand, both shuffle and AAFT surrogates (see Chapter 6) were different from chaotic model data. The fact that the shuffle surrogates, which are clearly random, produced the same forecasting signature as the neural data, complicates the interpretation and calls into question the earlier results.

These studies demonstrate a central point: no one approach should be relied upon in attempts to determine the type of dynamical behavior present in a given system. Judicious use should be made of all the computational tools at one's disposal, claims should be made carefully, and results should be reexamined periodically in the light of new developments.

References for Chapter 7

H Akaike (1979) A Bayesian extension of the minimum AIC procedure of autoregressive model fitting. *Biometrika* 66:237-242.

AM Albano, J Muench, C Schwartz, AI Mees, PE Rapp (1988) Singular-value decomposition and the Grassberger-Procaccia algorithm. *Physical Review A* 38:3017-3026.

GEP Box, GM Jenkins, GC Reinsel (1970) Time Series Analysis: Forecasting and Control. San Francisco: Holden-Day.

CC Canavier, SR Perla, PD Shepard (2004) Scaling of prediction error does not confirm chaotic dynamics underlying irregular firing using interspike intervals from midbrain dopamine neurons. *Neuroscience* 129:491-502.

M Casdagli (1989) Nonlinear prediction of chaotic time series. *Physica D* 35:335-356.

C Diks, JC van Houwelingen, F Takens, J DeGoede (1995) Reversibility as a criterion for discriminating time series. *Physics Letters A* 201:221-228.

JD Farmer, JJ Sidorowich (1987) Predicting chaotic time series. *Physical Review Letters* 59:845-848.

T Ikeguchi, K Aihara (1997) Difference correlation can distinguish deterministic chaos from $1/f^{\alpha}$-type colored noise. *Physical Review E* 55:2530-2538.

LP Lovejoy, PD Shepard, CC Canavier (2001) Apamin-induced irregular firing in vitro and irregular single-spike firing observed in vivo in dopamine neurons is chaotic. *Neuroscience* 104:829-840.

E Ott, T Sauer, JA Yorke (eds) (1994) Coping with Chaos: Analysis of Chaotic Data and the Exploitation of Chaotic Systems. New York: Wiley-Interscience.

DM Racicot, A Longtin (1997) Interspike interval attractors from chaotically driven neural models. *Physica D* 104:184-204.

PE Rapp (1993) Chaos in the neurosciences: cautionary tales from the frontier. *Biologist* 40:89-94.

T Sauer (1994) Reconstruction of dynamical systems from interspike intervals. *Physical Review Letters* 72:3811-3814.

CJ Stam, JPM Pijn, WS Pritchard (1998) Reliable detection of nonlinearity in experimental time series with strong periodic components. *Physica D* 112:361-380.

M Shelhamer (2005) Sequences of predictive saccades are correlated over a span of ~2 s and produce a fractal time series. *Journal of Neurophysiology* 93:2002-2011.

G Sugihara, RM May (1990) Nonlinear forecasting as a way of distinguishing chaos from measurement error in time series. *Nature* 344:734-741.

AA Tsonis, JB Elsner (1992) Nonlinear prediction as a way of distinguishing chaos from random fractal sequences. *Nature.* 358:217-220.

DJ Wales (1991) Calculating the rate of loss of information from chaotic time series by forecasting. *Nature* 350:485-488.

AS Weigend, NA Gershenfeld (eds) (1993) Time Series Prediction: Forecasting the Future and Understanding the Past. Santa Fe Institute Studies in the Science of Complexity, Vol. XV. Reading MA: Addison-Wesley.

Chapter 8

Recurrence analysis

Time-delay reconstruction allows us to perform computations and analyses on systems and their attractors in high-dimensional spaces. In many cases our intuition from three-dimensional space can serve well in envisioning attractors in higher dimensions, and the computations that we have seen so far indeed have straightforward extrapolations into higher dimensions. Nevertheless, the ability to grasp qualitatively the structure of an attractor in high-dimensional space is somewhat limited for most of us. This is where recurrence plots can help.

Originally developed as a qualitative method, recurrence plots reduce to two dimensions the features of higher-dimensional attractors, which can help us to visualize their structure. Shortly after their introduction, quantitative measures were introduced that are based on recurrence plots, leading to recurrence quantification analysis (RQA).

8.1 Concept and methodology

A recurrence plot is a graphical display of the spatial correlations in an attractor, in terms of time (Eckmann *et al.* 1987). It can give insight into periodicities, nonstationarity, and determinism in a signal.

To generate a recurrence plot, an attractor reconstruction is first performed via time-delay methods. Then, a reference point $x(i)$ is selected on the attractor, and a ball of radius r is centered on that point. If point $x(j)$ on the attractor is within this ball (that is, within distance r of $x(i)$), then a dot is placed at coordinates (i,j) on the recurrence plot. The procedure is repeated for all points $x(j)$, and the entire process repeated for all reference points $x(i)$. The recurrence plot has dimensions $N \times N$, where N is the number of points on the attractor. The coordinates (i, j) are the time indices of points on the attractor.

It is obvious from the construction what the recurrence plot tells us. If the attractor trajectory at two different times i and j comes close to itself – if the two points $x(i)$ and $x(j)$ are similar – then this is indicated by a dot at (i,j) on the recurrence plot. Thus any near-repetitions, or *recurrence*, of the attractor can be clearly visualized.

Typically, the threshold radius r is set so that there is a specified minimum number of neighbors $x(j)$ within each ball, although the radius can also be fixed at a given value or selected so as to obtain overall a certain percentage of "recurrent points" (i.e., to obtain a desired plot density). A more objective approach to setting the threshold distance (Matassini *et al.* 2002, Nichols *et al.* 2006) asserts that a count of the number of diagonal lines with "significant" length (more than three standard deviations above the mean length) should be unchanged as r increases, and recurrent points should appear primarily in these lines (rather than in random recurrences). This helps to set r above the noise level. Within this constraint, r should be small so that the recurrence procedure will be sensitive to changes in system dynamics.

Recurrence results can also be presented in a more general manner, with the graphical display of a *recurrence matrix*. Here, we plot the actual distances between pairs of points $(x(i), x(j))$ as a gray-scale (or colorized) image, with the gray level of each point (i,j) proportional to the distance $|x(i)-x(j)|$, rather than applying a threshold to the distances and plotting only those points which are within the distance threshold.

A theoretical underpinning for the use of recurrence matrices (McGuire *et al.* 1997) is based on the observation that the two-dimensional matrix of inter-point distances retains the unique topology (actually, geometry) of the original trajectories in state space. In other words, given a particular recurrence matrix, it is possible to reconstruct the state-space attractor. Knowledge of all inter-point distances serves to completely specify the attractor shape, no matter how high the embedding dimension in which the attractor exists. This is true except for a possible *isometric* transformation: translation or rotation – the attractor can be regenerated and it will have the correct shape, but it may be in a different place or oriented differently in the state space than the original.

It is not clear how this retention of geometry and dynamical states is affected by the thresholding that results in conventional recurrence plots.

8.2 Recurrence plots of simple systems

Examples of recurrence plots are shown in Fig. 8.2.1 for some simple systems. All plots are based on attractors reconstructed with embedding dimension $M=5$ and delay $L=5$. The threshold distance was set to provide plots with a density of 5% (5% of the $N \times N$ points are black dots). In each case, the abscissa and ordinate represent the time indices i and j, and a dot is placed at (i,j) if $|x(i)-x(j)|$ is less than a threshold distance; such points are said to be *recurrent*. All points along the main diagonals have dots on them, since each point is recurrent with itself. Since distance is commutative, the recurrence plot is symmetric about the main diagonal.

On the top left is a plot for Gaussian white noise ($N=1000$). There are small hints at structure in this plot (related to imperfections in the algorithm that generates the random numbers), but overall there are no obvious patterns. This is typical of uncorrelated noise.

On the top right is shown the result for a classic deterministic system: a pure sine wave ($N=200$). There are strong diagonal lines parallel to the main diagonal, spaced approximately 60 units apart. This represents the period of the sine wave; for a period of K points, $x(i)$ and $x(i+K)$ will coincide and therefore be recurrent. The lines are parallel to the main diagonal because, with period K, if $x(i)$ and $x(i+K)$ are recurrent, then so are $x(i+1)$ and $x(i+1+K)$, and so on as each path of the trajectory follows the one before, separated in time by K points. The diagonal lines are slightly thick because the threshold distance allows not only perfectly coinciding points, but also those nearly so, to be considered as recurrent.

To the lower left is a plot from the same sine wave, to which has been added noise and baseline drift which represents nonstationarity. The drift causes subsequent periods to become less recurrent, as reflected in shorter diagonal line segments at the upper left and lower right corners of the plot (where points maximally far apart in time are compared). Such nonstationarity in general is reflected in decreasing density as one moves away from the main diagonal. Additive noise contributes an overall "fuzziness."

The lower right plot is from the x variable of the Lorenz system ($N=1000$). As for the sine, deterministic structure leads to line segments parallel to the main diagonal. These segments are very short, reflecting rapid divergence of trajectories in the state space that come close to one another (a manifestation of sensitive dependence on initial conditions). Small regions of the plot resemble the recurrence plot from the sine; this is because over short time intervals the Lorenz system is approximately periodic (see lower left graph in Fig. 3.4.1b, and Fig. 4.4.1).

In Fig. 8.2.2 we present some specific features that appear in recurrence plots, and show how they arise from characteristics of the attractor in state space. All examples are drawn from small sections of attractors (trajectory segments) in two-dimensional embedding space ($M=2$) for ease in visualization. (Of course the true power of the recurrence plot stems from its ability to present information from attractors in higher-dimensional embedding spaces.)

The first example (top of Fig. 8.2.2.a) shows the presence of an isolated recurrent point in a random system. Nine points of a trajectory are shown. Points 2 and 6 (so labeled) are near each other, and so they are recurrent. Thus the recurrence plot shows isolated points at (2,6) and (6,2).

The next example shows two segments of an attractor trajectory, from time indices 1-11 and 21-27. (Although separated in the graph, these segments are to be considered as parts of a larger attractor and connected by the missing points 12-20.) As point 23 comes close to point 6, the two paths follow each other closely, and so the corresponding points are recurrent as each path progresses in time for two more steps. The fact that points 6-8 and 23-25 are recurrent is reflected by dots at the appropriate points on the recurrence plot (connected by lines to demonstrate the pattern more clearly). The resulting diagonal line segment, parallel to the main diagonal, is representative of recurrent paths. This in turn is suggestive of determinism, since points that are close together lead to time progressions that are similar, at least for short time spans.

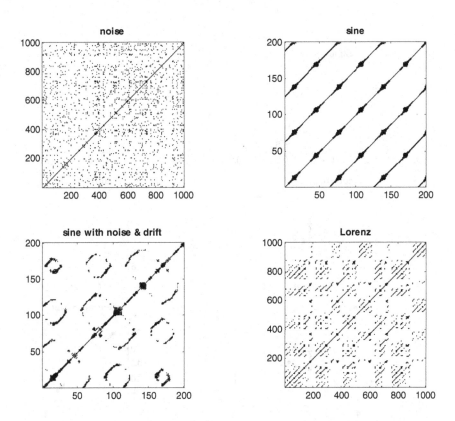

Figure 8.2.1. Recurrence plots from simple systems. Gaussian white noise (GWN, upper left) produces a recurrence plot with near-constant density at all locations; there are no obvious dynamical properties. A sine wave (upper right) produces strong recurrences parallel to the main diagonal, reflecting deterministic dynamics. The spacing of the diagonal lines reflects the periodicity of the sine wave. Adding noise and drift (a linear trend) to the sine (lower left) breaks up the strong periodic structure. The x variable of the Lorenz system (lower right) produces a rich structure that reflects deterministic dynamics, areas of near-periodicity, and rapid divergence of trajectories (reflected by the shortness of the diagonal line segments).

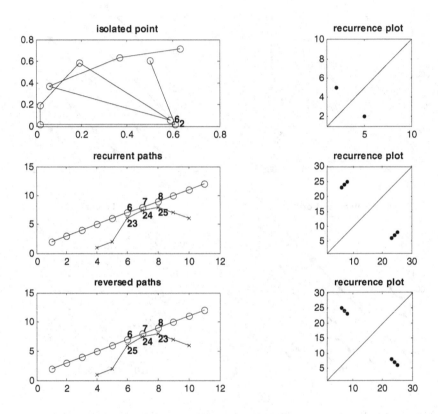

Figure 8.2.2a. Features of recurrence plots (right) and how they arise from features of the state-space attractor (left). Two trajectory segments that come close to each other at a single location produce an isolated recurrent point (top). If the two trajectory segments follow each other for a time, as is common for a deterministic system, diagonal lines result which are parallel to the main diagonal (center). If the two trajectory segments travel in opposite directions, then the diagonal lines are orthogonal to the main diagonal (bottom).

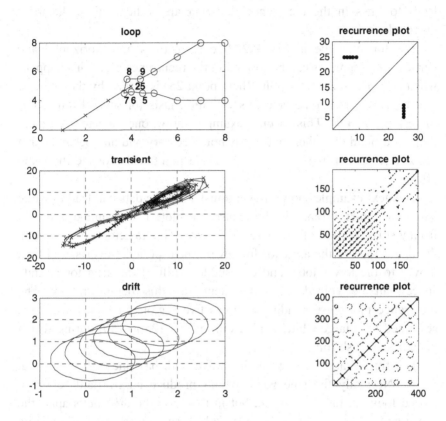

Figure 8.2.2b. Features of recurrence plots (right) and how they arise from features of the state-space attractor (left). A trajectory path that loops around an otherwise isolated point produces horizontal and vertical lines in the recurrence plot (top). A transient, or rapid change in dynamics, produces a gap in the recurrence structure (center); this is represented here by the shift between upper and lower lobes of the Lorenz attractor. An attractor that drifts in state space, for example due to nonstationarity, produces a recurrence plot that becomes less dense away from the main diagonal.

The bottom example in the figure is identical to the preceding one, except that the later trajectory segment is now reversed in time. This leads to lines in the recurrence plot that are orthogonal to the main diagonal.

The top example in Fig. 8.2.2b demonstrates how horizontal and vertical line segments can be formed in the recurrence plot. This happens when an isolated attractor point (here point 25) is visited by the attractor at a later or earlier time, and this visitation lasts for several time steps (here from 5 to 9). This specific example shows one trajectory segment looping around the other, but this is not necessary and any behavior that keeps one path in the vicinity of a single point will produce the same effect.

The next example shows how a transient in the data can lead to white bands of non-recurrence in the recurrence plot. This example is drawn from a small portion of the Lorenz attractor. The trajectory loops around the upper lobe of the attractor for several near-periodic cycles, and then moves to the lower lobe, and this sudden shift of the attractor in state space can be considered a transient for this demonstration. The recurrence plot correspondingly shows two regions of high recurrence, separated by a large white band (from indices 125 to 150) during which the attractor is shifting.

The final example shows a sine wave with baseline drift (as in Fig. 8.2.1). Normally the sine wave would produce an overlaid series of closed loops in the state space, but in this case the drift pulls apart the successive cycles. The recurrence plot, as before, shows periodic behavior that diminishes away from the main diagonal.

We can summarize these properties as follows:
- Isolated recurrent points can arise at random.
- Periodic structure is evident in repeating patterns.
- Diagonal line segments (parallel to the main diagonal) represent recurring paths, as in a deterministic system.
- Diagonals orthogonal to the main diagonal represent nearby paths that move in opposite directions with time.
- For chaotic systems, line segments are short due to rapid divergence of trajectories.

- Horizontal and vertical lines result from a path that is recurrent for several consecutive points with a single point at some other time, as in looping around a point.
- Nonstationarity or drift is reflected in decreasing density away from the main diagonal.
- A transient results in recurrence gaps.

While there is much value in this level of qualitative characterization (especially in comparing recurrence plots from different attractors), these properties can be quantified, as we show in the next section.

8.3 Recurrence quantification analysis (RQA)

Although essentially a qualitative method, steps have been taken to provide quantitative measures of some of the properties of the recurrence plot. This is termed *recurrence quantification analysis* (RQA: Webber & Zbilut 1994). The original implementation identified five measures derived from the recurrence plot:

1. Percent recurrence. This is simply the percentage of points that are defined as recurrent. Overall, periodic components should lead to more recurrent points. This alone can sometimes be a helpful discriminating statistic in comparing systems in different states. Of course care must be exercised in using comparable attractor reconstructions (*M*, *L*, *etc.*) and recurrence plot constructions (in particular selection of the threshold distance).
2. Percent determinism. This is the percentage of recurrent points that are in line segments parallel to the main diagonal. It is a normalized representation of the number of recurrent points that are derived from trajectory paths that follow each other closely, as in a deterministic system.
3. Entropy. This is a measure, based on information theory, of the "complexity" of a recurrence plot. It quantifies the distribution of the lengths of deterministic line segments (those parallel to the main diagonal). Shannon entropy is defined as:

$$E = -\sum_i P_i \log_2(P_i)$$

where P_i is the probability of a line segment having a length that falls into bin i of a histogram of segment lengths. Entropy is a measure of "uncertainty," in the sense that a more even distribution of segment lengths has higher entropy than a distribution in which most of the segment lengths are within a small range. (With the broader distribution we are less "certain" of the size range of any given segment, *a priori*.) Computation with base-2 logarithms leads to entropy in units of *bits*.
4. Ratio. This is the ratio of percent determinism to percent recurrence.
5. Trend. This quantifies any reduction in plot density away from the main diagonal, due to attractor drift. The percentage is found, of recurrent points that are in diagonals parallel to the main diagonal, as a function of the distance away from the main diagonal. The slope of this curve (percentage of points in diagonals vs. distance from main diagonal) is the trend measure. Values near zero indicate systems with little or no drift (nonstationarity).

Subsequent studies have added to this repertoire of quantitative measures. Several authors have pointed out that the lengths of the diagonal line segments are related to the time that it takes for trajectories to diverge. This is characterized mathematically by the *Lyapunov exponent*, which we do not cover in detail in this book because of the many subtleties in its extraction from experimental data.

Although we have identified diagonal line segments (those parallel to the main diagonal) as indicative of deterministic structure, in fact they can also occur due to temporal correlations in the data (e.g., Gao & Cai 2000). The recurrence in this case results from temporal proximity that leads to spatial proximity when the trajectory moves slowly, as when points are strongly correlated in time.

The presence of horizontal and vertical line segments in recurrence plots – as in the top example in Fig. 8.2.2b – has attracted considerable attention. Marwan *et al.* (2002) have defined *laminarity* as the proportion of recurrent points that are in these line segments, which they claim can reflect the presence of transitions from one chaotic state to another, during which an isolated part of the trajectory creates these segments when it comes close to an otherwise isolated point. In this sense, such points have also been termed *sojourn points* (Gao & Cai 2000), since they arise from a trajectory path passing near a point for a short time but not returning there on a regular basis (which would lead to periodic structures).

8.4 Extensions

Many extensions to the basic recurrence plot and RQA have been devised specifically to address transients and nonstationarities (Trulla *et al.* 1996, Gao 1999). Another extension is to use, instead of a threshold distance below which points are considered recurrent, a *range* of distances within which points are considered to be recurrent. In other words, if points are too close together in the state space, they are not considered recurrent. If the signal has not been oversampled in time, this method can be used to combat noise.

Too small an embedding dimension can lead to false nearest neighbors in the state space, as we saw in Chapter 4. Since recurrent points are close neighbors, a series of recurrence plots at increasing values of embedding dimension M can provide a diagnostic for determining if a proper embedding (attractor reconstruction) has been achieved. Similarly, Gao & Cai (2000) point out that line segments orthogonal to the main diagonal – as in the second example of Fig. 8.2.2a – can only occur if there has not been a proper embedding, because a true deterministic system cannot produce a trajectory that loops back on itself. This then can also be used to assess whether a proper embedding has been achieved.

Surrogate data sets, of course, can have great value when applied to recurrence plots, since in many cases it is possible actually to see the effects of different surrogate manipulations on the recurrence plot structure. Recurrence plots, with their ability to help visualize high-dimensional features, are perfect for this type of exploratory data analysis.

Finally, we note that extensions have been proposed to recurrence analysis for the investigation of two signals and their coupling (e.g., Zbilut *et al.* 1998, Marwan & Kurths 2002). This topic will be discussed in more detail in Chapter 9.

8.5 Examples

Recurrence plots are widely used, in a surprising range of life sciences applications: word occurrences in English (Webber & Zbilut 1996), respiration patterns (Webber & Zbilut 1994), neural spike trains (Kaluzny & Tarnecki 1993), cardiac dynamics (Zbilut et al. 1995), DNA coding (Xiao et al. 1993), and protein structure (Zbilut et al. 1998). Here we cover in some detail just one physiological application.

An early application of recurrence plots was to the detection of muscle fatigue in surface EMG recordings (Webber et al. 1995). With the forearm horizontal and the upper arm vertical, a 1.4 kg mass was placed in the hand of a human subject for 60 seconds, after which the mass was changed to 5.1 kg and maintained until task failure. Analysis consisted of examination of consecutive overlapping one-second EMG segments (sampled at 1 kHz).

Qualitatively, there is an organization of the recurrence plot into periodic structures late in the heavy loading epoch, relative to early in the light loading epoch. This is associated with an increase in the RQA measure of percent determinism (but not percent recurrence). To determine baseline values, 95% confidence intervals were obtained from a 60-second no-mass control condition. Deviations from this interval provide a basis for detecting physiological state changes due to muscle fatigue. Comparison of conventional linear and nonlinear measures was a central goal of the study. The measures of interest were f_c, the center frequency of spectral power in each time segment, and percent determinism in each segment.

Two major results came from this study. First, upon the transition from light to heavy arm loading, the attendant change in muscle state is detected by an increase in percent determinism in approximately half the time as it takes to detect by a decrease in f_c. Second, the change in percent determinism is approximately five times as great as the change in f_c. These results overall indicate that RQA measures can provide rapid and sensitive quantification of changes in physiological state.

References for Chapter 8

J-P Eckmann, S Oliffson Kamphorst, D Ruelle (1987) Recurrence plots of dynamical systems. *Europhysics Letters* 4:973-977.

JB Gao (1999) Recurrence time statistics for chaotic systems and their applications. *Physical Review Letters* 83:3178-3181.

P Kaluzny, R Tarnecki (1993) Recurrence plots of neuronal spike trains. *Biological Cybernetics* 68:527-534.

N Marwan, J Kurths (2002) Nonlinear analysis of bivariate data with cross recurrence plots. *Physics Letters A* 302:299-307.

N Marwan, N Wessel, U Meyerfeldt, A Schirdewan, J Kurths (2002) Recurrence-plot-based measures of complexity and their application to heart-rate-variability data. *Physical Review E* 66:26702-26709.

L Matassini, H Kantz, J Holyst, R Hegger (2002) Optimizing of recurrence plots for noise reduction. *Physical Review E* 65:021102-1:6.

JM Nichols, ST Trickey, M Seaver (2006) Damage detection using multivariate recurrence quantification analysis. *Mechanical Systems and Signal Processing* 20:421-437.

G McGuire, NB Azar, M Shelhamer (1997) Recurrence matrices and the preservation of dynamical properties. *Physics Letters A* 237:43-47.

LL Trulla, A Giuliani, JP Zbilut, CL Webber (1996) Recurrence quantification analysis of the logistic equation with transients. *Physics Letters A* 223:255-260.

CL Webber, MA Schmidt, JM Walsh (1995) Influence of isometric loading on biceps EMG dynamics as assessed by linear and nonlinear tools. *Journal of Applied Physiology* 78:814-822.

CL Webber, JP Zbilut (1994) Dynamical assessment of physiological systems and states using recurrence plot strategies. *Journal of Applied Physiology* 76:965-973.

CL Webber, JP Zbilut (1996) Assessing deterministic structures in physiological systems using recurrence plot strategies. In: MCK Khoo (ed) Bioengineering Approaches to Pulmonary Physiology and Medicine. New York: Plenum Press.

Y Xiao, Y Huang, M Li, R Xu, S Xiao (2003) Nonlinear analysis of correlations in Alu repeat sequences in DNA. *Physical Review E* 68:61913-61917.

JP Zbilut (2004) Unstable Singularities and Randomness: Their Importance in the Complexity of Physical, Biological and Social Sciences. New York: Elsevier Science.

JP Zbilut, A Giuliani, CL Webber (1998) Detecting deterministic signals in exceptionally noisy environments using cross-recurrence quantification. *Physics Letters A* 246:122-128.

JP Zbilut, A Giuliani, CL Webber, A Colosimo (1998) Recurrence quantification analysis in structure–function relationships of proteins: an overview of a general methodology applied to the case of TEM-1 β-lactamase. *Protein Engineering* 11:87-93.

JP Zbilut, M Zak, CL Webber (1995) Physiological singularities in respiratory and cardiac dynamics. *Chaos, Solitons & Fractals* 5:1509-1516.

Chapter 9

Tests for dynamical interdependence

A common problem in system analysis is the determination of coupling or interdependence between two measured quantities or two dynamical systems. Clearly this is an issue that arises in a great many cases in physiology, as for example in determining if a given signal produced by a system is functionally related to one of the system inputs.

Unfortunately, there is no single all-purpose measure – like the correlation coefficient for linear correlation – that can be applied to assess nonlinear relationships. Instead, several methods have been developed, centered on the question: what can knowledge of one signal tell us about the other signal? Mutual forecasting (prediction) is one such approach: can we use one signal to forecast future values of a different signal? If so, then they are very likely functionally related.

Another, somewhat more general, approach is to look for continuous mappings (topological transformations) between attractors generated from the two signals. This draws on the topological ideas presented in Chapter 4. The essence of the approach is that two points that are very close together on one signal's attractor should, through the nonlinear function that connects the two systems, lead to two points that are very close together on the other signal's attractor. This can be carried out in the reconstructed state space.

We now review these and similar approaches to this important problem. In this chapter we use $x(i)$ and $y(i)$, or $x_1(i)$ and $x_2(i)$, as two signals for which interdependence is to be assessed, or for their M-dimensional representations as points on attractors in state space. The interpretation as a time-series value or as an attractor point will be clear from the context. This nomenclature differs from that introduced in Chapter 4, where $x(i)$ is a time series and $y(i)$ are points on an attractor.

9.1 Concepts

The idea of coupling two chaotic systems so that their behaviors are identical might seem to be a logical contradiction, since a defining characteristic of chaos is that it is unpredictable (in a practical sense, in the long term). Nevertheless, there are conditions in which two identical chaotic systems can be coupled in this way. This situation has been studied extensively, as there is a direct practical application to the matter of secure communication over open channels (e.g., Cuomo & Oppenheim 1993).

Two categories of such *synchronization* between *identical* chaotic systems can be defined (Abarbanel 1996). First is the case in which one system *drives* the other. The second case is when the two systems are mutually coupled through some of their variables; the systems progress in time so as to reduce the differences between identical variables from each system. This is analogous to an optimal observer in linear systems theory (Kailath 1980).

A more general case – and one probably more relevant for physiology – is when we wish to detect coupling between two different systems. This more general condition is termed *generalized synchrony* (see below for a more restrictive definition). Here, inputs $u_1(i)$ and $u_2(i)$ produce outputs $y_1(i)$ and $y_2(i)$, respectively, from the two systems, and we expect there to be a function Φ that connects these two outputs: $y_1(i)=\Phi[y_2(i)]$. An important observation is that if this function exists, then $y_1(i)$ should be predictable based on knowledge of $y_2(i)$; the converse should be the case if Φ is invertible. Predictability (forecasting ability, as we termed it in Chapter 7) does not mean that we will necessarily know the function Φ, only that one signal can provide information to help forecast the behavior of the other.

Generalized synchrony was first defined by Rulkov *et al.* (1995) as a special case of interdependence in unidirectionally coupled systems. In their terminology, given a driver system **D** which produces x, and a response system **R** which produces y, we expect there to be a function Φ such that $y=\Phi(x)$. This represents a transformation from **D** to **R**. This is generalized synchronization if Φ is not a simple identity. The transformation Φ can be considered a topological transformation between the attractors from the systems **D** and **R**. Thus, aspects of continuity (smoothness) and differentiability come into play in determining if this transformation exists (as in determining the presence of a proper time-delay reconstruction in Chapter 4).

9.2 Mutual false nearest neighbors

The method of *Mutual False Nearest Neighbors* (MFNN) is an extension of the False Nearest Neighbors concept, presented in Chapter 4, for determining if the embedding dimension M is large enough for a proper time-delay attractor reconstruction. The underlying principle is that, for a "well-behaved" function Φ, points close together in the driving system $x(n)$ will lead to points that are close together in the response system $y(n)$. By "well-behaved" we mean that the function Φ is smooth, continuous, and invertible, so that points close together in the space of the driving system **D** should lead to points that are close together in the space of the response system **R**. The assumption is that Φ, if it exists, "preserves the identity of neighborhoods in state space" (Rulkov *et al.* 1995). (In this discussion we are referring to $x(n)$ and $y(n)$ as points on reconstructed attractors in state space.)

The basis of the procedure is the following. Choose a reference point $x(n)$ in the driving system. Find its nearest neighbor; let's say that it has time index *nnd*, so that the neighbor is $x(nnd)$. Now the point $y(n)$ from the response system, at the same time index n as $x(n)$, should have $y(nnd)$ as a near neighbor. (The closeness of the neighbors in x is preserved when transformed by Φ to y.) Likewise we can find, at the same time index n, point $y(n)$ in the response system **R**, and its nearest neighbor $y(nnr)$. Using the assumption that Φ is invertible, the points in the driving system that led to these points by way of the function Φ are then $x(n)$ and $x(nnr)$.

(It is useful to think of this procedure in terms of a limit process, so that as inter-point distances become smaller in **D** they become smaller in **R** as well. With a finite amount of discrete data, the best approximation to this limiting condition is the use of nearest neighbors.)

We can make a local linear approximation to the function Φ. In technical terms this is a Jacobian matrix $[\mathbf{D}]_{x(n)}$, but we can simply think of it as a linearized version of Φ that is valid for small distances between points. Then, for the points $x(n)$ and $x(nnd)$ in the driving system:

$[y(n)\text{-}y(nnd)] = \Phi(x(n))\text{-}\Phi(x(nnd)) \approx [\mathbf{D}]_{x(n)} [x(n)\text{-}x(nnd)]$.

Similarly, for the points $y(n)$ and $y(nnr)$ in the response system:

$[y(n)\text{-}y(nnr)] = \Phi(x(n))\text{-}\Phi(x(nnr)) \approx [\mathbf{D}]_{x(n)} [x(n)\text{-}x(nnr)]$.

Note that the Jacobian in both cases is an approximation taken at the point $x(n)$, although in one case it acts on points $x(n)$ and $x(nnd)$ which are close in the driving system **D**, while in the other case it acts on points $x(n)$ and $x(nnr)$ which are the points that project to $y(n)$ and $y(nnr)$ which are close in the response system **R**.

In each case, $[\mathbf{D}_{x(n)}]$ is the same approximation matrix. Rearranging these two equations to form expressions for $[\mathbf{D}_{x(n)}]$, and taking the ratio of the magnitudes of the two expressions for $[\mathbf{D}_{x(n)}]$, leads to:

$$\frac{|y(n) - y(nnd)| \, |x(n) - x(nnr)|}{|x(n) - x(nnd)| \, |y(n) - y(nnr)|}$$

This expression, since it is the ratio of two quantities that each define $[\mathbf{D}_{x(n)}]$, should be close to the value 1.0 if the function Φ exists.

A problem arises with this preliminary formulation. As embedding dimension increases, the average distance between nearest neighbors increases. This increases the magnitude of the denominator in the expression above, which can make the ratio artificially small. Also, as embedding dimension increases, nearest neighbors are not necessarily close together anymore, as we saw with the original False Nearest Neighbors method in Chapter 4.

A revised version of this criterion ratio is created as follows. The response $y(n)$ is reconstructed in space $\mathbf{R}_E$ of dimension M_R, and this dimension is held constant. (M_R must be sufficiently large for the reconstruction to be a proper embedding as described in Chapter 4.) Next we form a reconstruction of the driving system $x(n)$ in space $\mathbf{D}_E$ with dimension M_D. Pick a value of embedding dimension for the driver system: M_D. Choose a point $x(n)$ and find its nearest neighbor $x(nnd)$ for the driver; this is done in the space of dimension M_D, which may be too small, in which case these points are false neighbors. Next, find the values of these points after passing through the function Φ; these are $y(n)$ and $y(nnd)$, and again if $x(n)$ and $x(nnd)$ are not really neighbors then these will not be neighbors either. Finally, find the nearest neighbor $y(nnr)$ of the point $y(n)$ for the response, in space M_R.

$$R_{FNN} = \frac{|y(n) - y(nnd)|}{|y(n) - y(nnr)|}$$

defines a criterion for MFNN.

If there is no synchronization, then Φ does not exist (or does not preserve local neighborhoods as in our definition). In this case, no matter how large we make M_D, the fact that $x(n)$ and $x(nnd)$ are close together does *not* necessarily mean that $y(n)$ and $y(nnd)$ are close together. On the other hand, since M_R is sufficiently large for a proper reconstruction of the response system $y(n)$, the points $y(n)$ and $y(nnr)$ should be *actual* nearest neighbors and therefore close together. The numerator is large while the denominator is small, and R_{FNN} will be large for all M_D.

If there is synchronization, then Φ exists and preserves local neighborhoods. For small values of M_D, $x(n)$ and $x(nnd)$ will be false neighbors, and the argument above applies: the ratio R_{FNN} is large. As M_D increases to the point where there is a proper reconstruction of all attractors, the fact that $x(n)$ and $x(nnd)$ are close together now *does* imply that $y(n)$ and $y(nnd)$ are also close together, and so the numerator decreases in value, and the ratio decreases as well. Thus a criterion for the existence of Φ is that the ratio above decreases with increasing M_D.

Distances in the MFNN equation are computed in the space with dimension M_R, no matter how large M_D becomes. (However, nearest neighbors $x(n)$ and $x(nnd)$ are found in the space M_D.) This avoids the problem of near neighbors becoming artificially far apart in large embedding spaces.

This ratio can be normalized to compensate for the fact that points will on average become farther apart as M_D increases. This is done by forming the same ratio, using points from the driving system $x(n)$:

$$D_{FNN} = \frac{|x(n) - x(nnd)|}{|x(n) - x(nnr)|}.$$

The final MFNN parameter is the ratio of these two ratios:

$$P(n, M_R, M_D) = \frac{R_{FNN}}{D_{FNN}} = \frac{|y(n) - y(nnd)| \, |x(n) - x(nnr)|}{|y(n) - y(nnr)| \, |x(n) - x(nnd)|}.$$

As noted, the behavior of this parameter as a function of M_D is a criterion to assess the existence of the function Φ. If Φ does exist, then the value will be large when M_D is small, and will be approximately equal to one as M_D increases.

(Note that the equations above are correct in Rulkov et al. 1995 but incorrect in Abarbanel 1996. Also, the original formulation used squared distances to reduce the computational burden of taking square roots.)

We note that the limiting process implicit in the derivation above can be made more explicit, and this is this basis for one method of detecting determinism in a single time series (Kaplan 1994); we will see this in Chapter 11. One should keep in mind that there is an asymmetry in the MFNN procedure, due to the fact that nearest neighbors in the driving system are determined at increasing embedding dimension M_D, while the embedding dimension M_R of the response system is held constant. This establishes the nature of the driver-response relationship and makes this implicitly a test for unidirectional coupling.

9.3 Mutual prediction, cross-prediction

Another approach to the problem of nonlinear coupling is based on the ability to use information from one system to forecast the future course of the other system. This is *mutual forecasting or prediction* (Schiff et al. 1996). (Although the term "prediction" is most often used for the procedure that we have termed "forecasting" in Chapter 7, we retain the latter term here also and refer to "mutual forecasting," even though it is also known as "mutual prediction.") As in the previous section, $x(n)$ and $y(n)$ here represent points on reconstructed attractors.

Begin by normalizing each time series by subtracting its mean and dividing by its standard deviation (this makes each time series zero-mean with unit variance).

In conventional forecasting, nearest neighbors of a reference point in a single time series are used to predict the future course of that reference point (see Chapter 7). In mutual forecasting, a reference point in one system is found (call it $x(n)$), and the corresponding point at the same time is identified in the other system ($y(n)$). The nearest neighbors to the point $y(n)$ are found; these points are $y(nny1)$, $y(nny2)$, and so on. These same times indices are now used to identify points in x: $x(yyn1)$, $x(yyn2)$, and so on. The future course of these nearest neighbors ($y(nny1)$, $y(nny2)$,...) is used to forecast the future course of $x(n)$.

It is perhaps better to think of a reference *time* being identified (n). We wish to forecast from $x(n)$ using information in the $y(i)$ attractor. So we find the neighbors of $y(n)$, which is at the reference time n but on the y attractor. As with MFNN, if the systems are coupled via a function Φ that preserves neighborhoods, then the time indices of these neighbors of $y(n)$ should also be the time indices of the neighbors of $x(n)$. So we use these time indices (derived from nearest neighbors in y) to select points in x, and these points are used to forecast the time course of $x(n)$.

As in Chapter 7, either the forecasting error or the correlation between actual and forecast values can be used to quantify the quality of the forecasting. It is common, when using forecasting error, to normalize it, for example by dividing by the error that would be obtained if every forecast value was simply the mean of the data.

With two time series we can produce four forecasts. Two of them are conventional forecasts as in Chapter 7: forecast $x(i)$ based on $x(i)$ alone, and forecast $y(i)$ based on $y(i)$ alone. The other two are mutual forecasts: forecast $x(i)$ based on $y(i)$ as above, and forecast $y(i)$ based on $x(i)$.

Generalized synchrony, as defined previously, implies that systems X and Y will exhibit mutual forecasting in each direction (X→Y and Y→X). This is because the function Φ is assumed to be smooth and invertible, and the forecasting procedures do not stipulate a direction of coupling (information flow). However, bidirectional mutual forecasting does not imply the presence of generalized synchrony (existence of Φ) unless the systems are coupled unidirectionally. In other words, bidirectional coupling can lead to bidirectional mutual forecasting even without generalized synchrony. This is because, with bidirectional coupling, either of the systems X or Y contains all information about both systems. Recall that unidirectionally coupled systems exhibit generalized synchrony if the "well-behaved" function Φ exists and $Y=\Phi(X)$. On the other hand, two systems can be bidirectionally coupled, with information flowing between them in both directions (the dynamics of X depend on the values of y and the dynamics of Y depend on the values of x). In this case, each system contains information about the other, but the connection between them is not necessarily smooth and invertible as in generalized synchrony. Again, for generalized synchrony one must show the existence of the transformation Φ. If we know that the systems are unidirectionally coupled, so that information flows in only one direction (X→Y for example), then bidirectional mutual forecasting does imply the presence of generalized synchrony.

In the case of unidirectional coupling between systems X and Y, we will be able to forecast $x(i)$ from $y(i)$ if X drives Y. This is because the dynamics of X are embedded in Y, so $y(i)$ has information on $x(i)$ and can be used to forecast $x(i)$. Not until the coupling is strong enough to establish generalized synchrony between the two systems (if this does happen) will there be mutual forecasting in both directions (X forecasts Y and vice versa). In this way, if we know in advance that there is unidirectional coupling (as in many neural systems, for example), then mutual forecasting can be used to determine the direction of coupling.

An example of the use of mutual forecasting is found in the three accompanying figures. Following Schiff *et al.* (1996) we examine two coupled Hénon systems (see section 9.6 for a description of this data set). In each graph, normalized forecasting error is plotted as a function of the number of time steps into the future over which the forecast is carried out. The error that would result from simply using the mean of the time series for each forecast value is equal to 1.0; values less than this indicate the ability to forecast future values (determinism). Forecasting is carried out by finding the 10 nearest neighbors of each reference point, and then finding the mean of the projections of these neighbors from one to ten steps into the future.

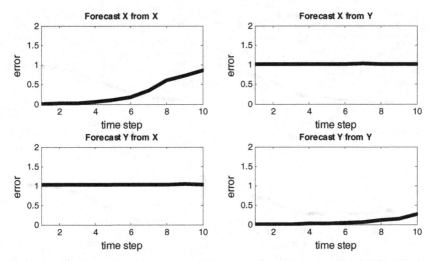

Figure 9.3.1. Mutual forecasting of two Hénon systems, with no coupling ($C=0$). When each system is used to forecast itself, low error at small forecasting steps demonstrates determinism in each system. High error when forecasting from one system to the other demonstrates lack of coupling between them.

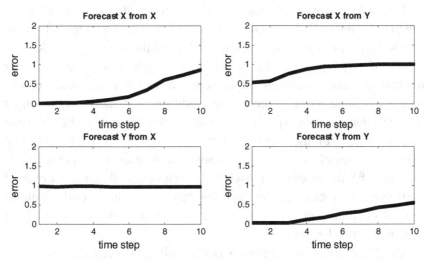

Figure 9.3.2. Mutual forecasting of coupled Hénon systems, with low coupling ($C=0.1$). Ability to forecast X from Y (initial low error) demonstrates presence of X dynamics in Y, while lack of forecasting of Y from X (high error) shows that X contains little information on the dynamics of Y. This indicates unidirectional coupling.

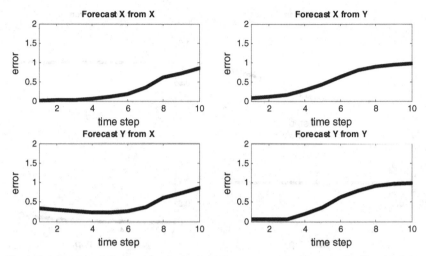

Figure 9.3.3. Mutual forecasting of coupled Hénon systems, with high coupling (C=0.9). Ability to forecast both X from Y and Y from X demonstrates generalized synchrony in this system, which we know from construction is unidirectionally coupled.

Fig. 9.3.1. shows forecasting for the two systems when they are completely uncoupled (coupling parameter C=0). The top left graph shows the result of forecasting future values of the driver system X from values of $x(i)$ only, and the bottom right graph shows the result for forecasting future values of the response system Y from values of $y(i)$ only. Low initial error in these cases indicates that each system, on its own, has deterministic dynamics. There is complete lack of the ability to forecast values of the driver X from the response Y (top right) and likewise values of the response Y cannot be forecast from the driver X (bottom left): the forecasts in each case are no better than those obtained from using the mean value as each forecast value. These results indicate lack of coupling between the systems; although each is chaotic, their dynamics are independent.

Fig. 9.3.2 shows forecasting when coupling between the systems is small (C=0.1); there is not yet generalized synchrony, so the coupling is unidirectional (information flows from driver X to response Y). Now there is some ability to forecast the future values of the driver X from values of the response Y, as shown by a reduction in the error (top right).

This demonstrates the unidirectional coupling: information on the dynamics of X appears in the response Y and can be used to forecast X. Lack of forecasting in the opposite direction is a result of the unidirectional flow of information.

Finally, Fig. 9.3.3 shows forecasting when coupling is strong (C=0.9). In this case we have the ability to forecast in both directions (X from Y and Y from X), which tells us that the dynamics of each system are present in the other. We know from construction of the system that the flow of information is unidirectional (X→Y). Therefore the dynamics have become "entrained": generalized synchrony.

9.4 Cross-recurrence, joint recurrence

One of the first uses of recurrence plots (see Chapter 8) to examine the dynamical relationship between two signals was in a signal-processing context, as a way to detect a signal contaminated in noise (Zbilut et al. 1998). A probe signal is required, which resembles the noise-contaminated signal. This resemblance can be quite loose, however, as a square wave and a sine wave of different frequencies will exhibit some cross-recurrence, which can be used to identify one of the signals when hidden in noise. In this approach it is not always necessary to perform an embedding or time-delay reconstruction before the recurrences are found; distances can simply be measured between values in each time series as $d(i,j)=|x_1(i)-x_2(j)|$, and recurrence points for which $d(i,j)<d_{threshold}$ plotted at point (i,j). The percentage of recurrent points is a measure of similarity between x_1 and x_2. Note that, in general, cross-recurrence plots (CRP) are not symmetric. An example of a CRP is presented in Fig. 9.4.1, which shows cross-recurrence between the Lorenz x and y variables. This methodology was extended and generalized in later work (Marwan & Kurths 2002).

Although a few quantitative parameters have been defined from these plots, this remains largely a qualitative approach. One shortcoming is that the two signals must "overlay" each other in the state space; even a simple shift (position offset) in one signal will move them apart and destroy the recurrences. Appropriate normalization of the signals can help, but for example if two attractors are deterministically related but rotated relative to each other in the state space, their cross-recurrences will be poor. This is not as big a drawback if the two time series are examined directly, without embedding or reconstruction.

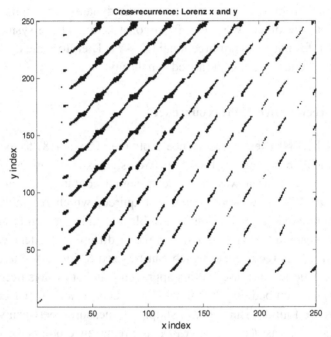

Figure 9.4.1. Example of cross-recurrence between the x and y variables of the Lorenz system.

A more refined mechanism for examining multi-variable recurrences is the *joint recurrence plot* (Romano *et al.* 2004). In this method, we begin by creating separate recurrence plots for the two variables in question, call them x_1 and x_2. Then the joint recurrence plot is formed by placing a dot at location (i,j) whenever there is a dot at that location in the recurrence plots for *both x_1 and x_2*. In other words, the joint recurrence plot identifies pairs of time indices at which the two signals are simultaneously (jointly) recurrent within themselves.

Mathematical construction of a joint recurrence plot is quite simple. Recall from Chapter 8 that the recurrence matrix for a given system contains all inter-point distances. A conventional recurrence plot is created by applying a threshold to these distances, which can be thought of as modifying the recurrence matrix to have only the values 1 (where the distance is less than the threshold) and 0 (otherwise); call this a *recurrence plot matrix (RP)*. The joint recurrence plot matrix is created by multiplying the RP for x_1 by the RP for x_2, on an element-by-element basis (i.e., not typical matrix multiplication). This places a 1 in the joint matrix whenever there is a 1 in both single-variable matrices at the same location.

There is another approach to multi-variable recurrence (Nichols *et al.* 2006), which might be termed a *multivariate recurrence plot*. It is created by concatenating the different signals to be analyzed ($x_1(1...N_1)$ and $x_2(1...N_2)$), as described in section 4.7:

$z(i)=[\ x_1(1+(i-1)J_1),\ x_1(1+(i-1)J_1+L_1),\ ...,\ x_1(1+(i-1)J_1+(M_1-1)L_1)\ ...$
$x_2(1+(i-1)J_2),\ x_2(1+(i-1)J_2+L_2),\ ...,\ x_2(1+(i-1)J_2+(M_2-1)L_2)\ ...\ ]$.

The multivariate recurrence plot is created from this signal. Note that the lower left $N_1 \times N_1$ section of such a plot is just the single-variable recurrence plot for x_1, and the upper right $N_2 \times N_2$ section is just the single-variable recurrence plot for x_2, so only one of the other sections of the plot needs to be displayed to show the mutual recurrences. This procedure can be easily extended to more than two variables, and in general each can be treated with a different set of embedding parameters M, L, and J.

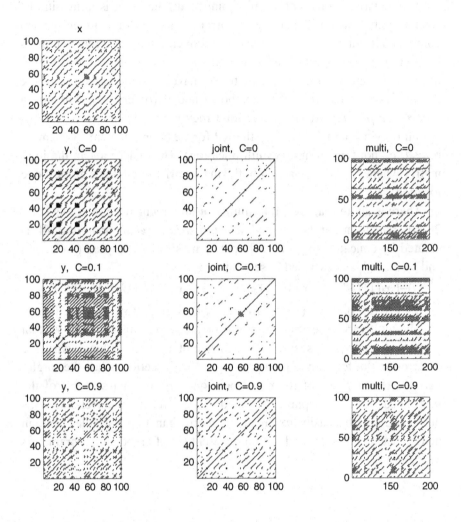

Figure 9.4.2. Examples of joint (center column) and multivariate (right column) recurrence plots. Data are from two Hénon systems with different degrees of coupling, increasing from row two to row four. The top row shows the conventional recurrence plot for the x variable from one Hénon system.

Examples of these two recurrence methods are provided in Fig. 9.4.2. The data sets are from two coupled Hénon systems as in section 9.3 (data described in section 9.6). The top recurrence plot in the figure is from 100 points of the Hénon x variable; the diagonal lines indicate strong determinism. Each subsequent row in the figure contains results for different degrees of coupling between the two Hénon systems: $C=0$ (second row), $C=0.1$ (third row), and $C=0.9$ (bottom row). (This is analogous to Figs. 9.3.1, 9.3.2, and 9.3.3 above.) In each row, the recurrence plot for the associated Hénon variable from the second system (call it y) is to the left, the joint recurrence plot is in the center, and the multivariate recurrence plot is to the right.

As coupling strength increases, the number and average length of the diagonal lines in the joint recurrence plots increase as well, and there are relatively fewer isolated recurrent points (random joint recurrences become less common relative to deterministic joint recurrences). The multivariate recurrence plots are less easily interpreted, although generally the diagonal line segments become more prevalent as coupling increases.

9.5 Mathematical properties of mappings

The specific procedures described in this chapter are in large part practical implementations of concepts of continuity presented in more detail by Pecora *et al.* (1995). We will see some of these concepts again in Chapter 11 in the context of detecting determinism in a single time series. Here we review some of the main ideas.

A straightforward approach to identifying a continuous nonlinear transformation between two time series is based on the concept that nearby points in one system (X) should not be transformed or mapped to points that are far apart in the other system (Y). We can test this by finding those points $x(nnx_i)$ within distance δ of a reference point $x(i)$ from system X, and those points $y(nny_i)$ within distance ε of a reference point $y(i)$ from system Y; the reference point has the same time index i in each case. In the case of continuity, all the $y(nny_i)$ should be within ε of $y(i)$; we decrease δ until this is true. Given a total of N points on the attractor, N_δ is number of points close to $x(i)$, and N_ε is number of points close to $y(i)$. (The points $x(i)$ and $y(i)$ themselves are not counted.) In general, N_ε might be greater than N_δ, since points not close to $x(i)$ may still map to points that are close to $y(i)$.

Under the null hypothesis of a random distribution of the $y(i)$, the probability that all of the δ-neighbors in X will map to within ε of $y(i)$ is just the likelihood that any randomly selected set of N_ε points, out of a total of N points, would do so: $P(N_\delta)=(N_\varepsilon/N)^{N\delta}$. If this probability (of all N_δ points mapping close to $y(i)$ by pure chance) is small enough, then we can reject the null hypothesis, which implies (but does not prove) that the mapping from X to Y is continuous. In practice, a normalization of this probability value is suggested based on the maximum possible probability value that could be obtained from a binomial distribution under the null hypothesis. This method was developed and tested more fully by Netoff *et al.* (2004).

Inverse continuity is assessed in the same way, reversing the roles of the X and Y systems. The next important property of a mapping (specifically a homeomorphism, see Chapter 4) is injectivity. That is, the mapping function should be one-to-one: a single value of $x(i)$ should lead to only one value of $y(i)$. This can be assessed by determining if the mapping function and its inverse are both continuous. If they are, then we have some confidence that the function is one-to-one. The continuity criteria above can be used for this test.

Other features of a "well-behaved" function or mapping from X to Y are differentiability and preservation of dimension (rank invariance). These concepts and their assessment are more intricate and will be omitted here. Again the work by Pecora *et al.* (1995) can be consulted for details.

9.6 Multivariate surrogates and other test data

In section 6.9 of Chapter 6 is described one type of multivariate surrogate data, which could be applied to verify algorithms for nonlinear interdependence such as those presented above.

It is also possible to create coupled data sets from systems with known dynamics. The examples presented in sections 9.3 and 9.4 made use of data from two coupled Hénon systems (Schiff et al. 1996). This system (Hénon 1976) is a two-dimensional discrete-time mapping, and is a commonly used example system, much like the continuous-time Lorenz system that we have seen several times. The Hénon system is known to be chaotic for some parameter values.

A set of coupled Hénon systems is:

driver X: $x(i+1) = 1.4 - x^2(i) + 0.3u(i)$

$u(i+1) = x(i)$

response Y: $y(i+1) = 1.4 - [Cx(i) + (1-C)y(i)]y(i) + Bv(i)$

$v(i+1) = y(i)$

The top set of equations, for the driver system X, is a conventional Hénon system, which maps two-dimensional points $(x(i), u(i))$ from one generation i to the next. In the coupled systems, the response system Y is identical to the X system if parameter B is 0.3 and the coupling parameter C is zero. The value of B can be changed to examine coupling of non-identical systems. When C is non-zero, some of the dynamics from X are introduced into the Y system, leading to unidirectional coupling. Similar manipulations can be made with other chaotic systems (e.g., Rulkov et al. 1995). Such coupled data sets provide excellent test cases for algorithms that are meant to assess nonlinear coupling and interdependence.

9.7 Examples

Several studies have examined the dynamics of human EEG data, especially with respect to the detection of imminent epileptic seizures. We will review some of this material in a later chapter. Here, we present the outline of one such study, which used mutual forecasting to look for interactions between different cortical areas during normal EEG activity (Breakspear & Terry 2002).

This particular study used a mutual forecasting procedure that is very much like that described in section 9.3, with some simple modifications.

The False Nearest Neighbors algorithm (section 4.4) was used to set the embedding dimension M, and instead of nearest neighbors, each forecast was based on a simplex of $2M$ points with the contributions of the points in the simplex weighted inversely to their distances from the reference point (Sugihara & May 1990, also see Chapter 7). Also, instead of comparing forecasts to those formed from the mean of the time series (as above), random forecasts were generated based on a random set of points and their distances from a randomly selected reference point.

This algorithm was then applied to multichannel EEG data. Specifically, coupling of EEG activity between cerebral hemispheres was assessed by analysis of data taken at the same time from electrode pairs on the right and left sides of the head. Forecasting results were compared to those from bivariate surrogate data sets which preserve the amplitude distribution, power spectra, and cross-spectra; comparison with 19 samples of each surrogate provided a significance level of 0.05 (see Chapter 7). Epochs of 2.048 sec were extracted from 130 sec trials with the eyes open and 130 sec with the eyes closed, from 40 normal subjects. In approximately 3% of eyes-open epochs and 5% of eyes-closed epochs, there was some nonlinear interdependence (ability to forecast data from one hemisphere using data from the other). Far fewer epochs showed mutual forecasting in both directions.

This finding led to a more detailed investigation of those epochs in which nonlinear interdependence was identified. It was found that there is a tendency for the phases of different frequency components to synchronize, across the two hemispheres, in the interdependent epochs. This cross-frequency effect is a nonlinear phenomenon (see Chapter 3). The finding of nonlinear coupling was used to categorize the data for further analysis, which revealed in more detail the characteristics of the interdependence.

References for Chapter 9

The material in section 9.2 is drawn heavily from Rulkov *et al.* (1995) and Abarbanel (1996), and that in section 9.3 is drawn heavily from Schiff *et al.* (1996).

HDI Abarbanel (1996) Analysis of Observed Chaotic Data. New York: Springer-Verlag. Chapter 9.

M Breakspear, JR Terry (2002) Detection and description of non-linear interdependence in normal multichannel human EEG data. *Clinical Neurophysiology* 113:735-753.

KM Cuomo, AV Oppenheim (1993) Circuit implementation of synchronized chaos with applications to communication. *Physical Review Letters* 71: 65-68.

M Hénon (1976) A two-dimensional mapping with a strange attractor. *Communications in Mathematical Physics* 50: 69-77.

T Kailath (1980) Linear Systems. Englewood Cliffs, NJ: Prentice-Hall.

DT Kaplan (1994) Exceptional events as evidence for determinism. *Physica D* 73:38-48.

N Marwan, J Kurths (2002) Nonlinear analysis of bivariate data with cross recurrence plots. *Physics Letters A* 302:299-307.

TI Netoff, LM Pecora, SJ Schiff (2004) Analytical coupling detection in the presence of noise and nonlinearity. *Physical Review E* 69:017201-1:4.

JM Nichols, ST Trickey, M Seaver (2006) Damage detection using multivariate recurrence quantification analysis. *Mechanical Systems and Signal Processing* 20:421-437.

LM Pecora, TL Carroll, JF Heagy (1995) Statistics for mathematical properties of maps between time series embeddings. *Physical Review E* 52:3420-3439.

MC Romano, M Thiel, J Kurths, W von Bloh (2004) Multivariate recurrence plots. *Physics Letters A* 330:214-223.

NF Rulkov, MM Sushchik, LS Tsimring, HDI Abarbanel (1995) Generalized synchronization of chaos in directionally coupled chaotic systems. *Physical Review E* 51:980-994.

SJ Schiff, P So, T Chang, RE Burke, T Sauer (1996) Detecting dynamical interdependence and generalized synchrony through mutual prediction in a neural ensemble. *Physical Review E* 54:6708-6724.

G Sugihara, RM May (1990) Nonlinear forecasting as a way of distinguishing chaos from measurement error in time series. *Nature* 344:734-741.

JR Terry, M Breakspear (2003) An improved algorithm for the detection of dynamical interdependence in bivariate time-series. *Biological Cybernetics* 88:129-136.

JP Zbilut (2004) Unstable Singularities and Randomness: Their Importance in the Complexity of Physical, Biological and Social Sciences. New York: Elsevier Science.

JP Zbilut, A Giuliani, CL Webber (1998) Detecting deterministic signals in exceptionally noisy environments using cross-recurrence quantification. *Physics Letters A* 246:122-128.

Chapter 10

Unstable periodic orbits

This and the next three chapters will deal, briefly, with subjects that are slightly off the main track of our exposition. These can be considered advanced topics and, in some cases, represent methods that have not seen as much widespread use as those in the preceding chapters.

10.1 Concepts

In the state space, the trajectories of a deterministic system, even a chaotic one, are not haphazard. Recall that a chaotic system is deterministic but aperiodic – the trajectory never repeats itself. Yet the (potentially) infinitely long trajectory occupies only a finite portion of the state space. This leads to the fractal nature of the strange attractor of a chaotic system. (These concepts were discussed extensively in previous chapters.)

Another aspect of these geometric characteristics is that the state-space attractor can often be decomposed into a set of almost-periodic orbits, where the trajectory almost returns to a previous location. Such orbits are unstable, meaning that points that are not on a given orbit are *not* attracted to that orbit, and the orbit does not last forever but rather dissipates.

While recognizing that actual state-space orbits may be only approximately periodic, we nevertheless will refer to them as *unstable periodic orbits*. Thus we can think of a given attractor as being composed of a substrate of such periodic orbits, each one unstable. Furthermore, there is a hierarchy of these orbits with increasingly longer periods, as we will see in the next section.

10.2 Example

Here we follow the example of Lathrop and Kostelich (1989) to demonstrate the concepts of periodic orbits in experimental data. The data set is from the Belousov-Zhabotinskii chemical reaction (see Chapter 4). In Fig. 10.2.1, at the upper left, 7000 points from this data series are plotted as a 2-dimensional time-delay reconstruction of the attractor. (Obviously, to avoid trajectory crossings, the actual state space must have dimension of at least three.) The other panels in this graph each show a single example of a near-periodic orbit. Periodicity is defined as the trajectory returning to one of its previous values, within some small tolerance. This can be seen by the small gap in some of the plots, where the trajectory has almost but not quite returned to its earlier location. The *period 1* orbit is approximately 125 points long. Applying the same criterion to identify near-recurrence (when the trajectory has returned close to one of its previous locations) leads to the identification of other orbits of higher periodicity; orbits of *period 2* (~250 points) and *period 3* (~375 points) are shown. (In the cases of period 2 and period 3, although it appears that the trajectory revisits itself at an earlier time than that used to delineate the orbit, these apparent trajectory crossings are due to the fact that the high-dimensional attractor is being viewed in a two-dimensional projection.)

These orbits can be characterized more completely by setting a threshold to determine when the trajectory has returned (almost) to its previous location. Each point on the attractor was selected, in sequence, as a starting point, and the distance of every point in time beyond the starting point was determined, until a point was found that was within the threshold distance of the starting point. This identified a near-periodic orbit, and the time interval between the starting point and the return point is the period. These periods are plotted as a histogram in Fig. 10.2.2. This simple analysis reveals the existence of a number of orbits with periods that are nearly multiples of the period-1 orbit of approximately 125 points.

In practice, after identifying these orbits, their stability properties are usually assessed. This is accomplished by determining, each time the orbit completes one period, if the resulting periodic points are drawn closer together (indicating a stable orbit) or thrown farther apart (indicating instability). It is also possible, in some cases, to extract dimensions and other measures from the properties of the periodic orbits (Auerbach *et al.* 1987).

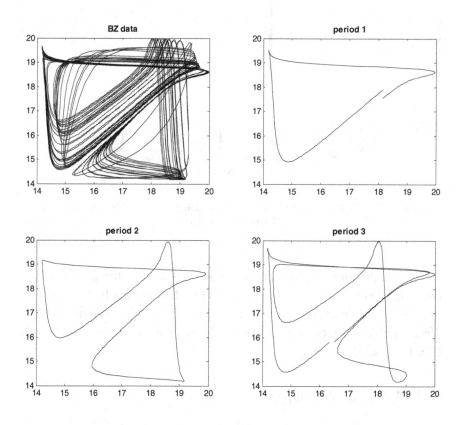

Figure 10.2.1. State-space attractor reconstructed with data from the Belousov-Zhabotinskii chemical reaction, showing unstable periodic orbits of various periods.

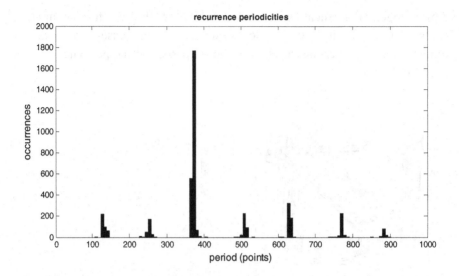

Figure 10.2.2. Distribution of the periods of unstable periodic orbits in the Belousov-Zhabotinskii attractor.

A very practical application of periodic orbits is in the area of *chaos control*, where small perturbations to the system are applied in order to move the state onto an orbit with the desired properties (Garfinkel *et al.* 1992). There is a large and growing volume of literature on this subject. (See Chapter 12.)

10.3 Physiological examples

The majority of recent practical applications of the concept of periodic orbits make use of a data transformation that groups the data closer to the periodic points (So *et al.* 1996). Making use of this transformation, a related study found periodic orbits (an indicator of deterministic dynamics) in neural spike intervals from rat hippocampus and in human EEG (So *et al.* 1998).

Following this same path, another study (Le Van Quyen *et al.* 1997) similarly found unstable periodic orbits (UPOs) in human EEG in a patient during ongoing focal epileptic activity. After presentation of a visual stimulus, there was a change from one to three or four periodic orbits. This shows the influence of cognition on neural dynamics even within epileptic activity, and confirms that collective neural dynamics organize into a "coherent oscillatory state" due to processing of a sensory stimulus (Skarda & Freeman 1987, see also discussion in Chapter 1). These findings of UPOs and increasingly organized neural dynamics during epileptic activity form the basis for *chaos control* approaches as a therapeutic intervention for epilepsy (see Chapters 12 and 18).

A study that did not make use of this data transformation, but instead used the simple recurrence approach of section 10.2 to identify periodic orbits, investigated human cardiac rhythms (Narayanan *et al.* 1998). Here again, there was an increase in the number of UPOs from healthy to pathological, increasing from 3-4 to 10 or more in the case of ventricular fibrillation. This measure in some cases demonstrated a change with pathology when other measures such as dimension and spectrum did not. The number and distribution of UPOs thus might provide a characteristic signature of different pathological conditions.

We note that the use of surrogate data (Chapter 6) is essential in these studies in order to provide a measure of statistical reliability.

References for Chapter 10

D Auerbach, P Cvitanović, J-P Eckmann, G Gunaratne, I Procaccia (1987) Exploring chaotic motion through periodic orbits. *Physical Review Letters* 58:2387-2389.

A Garfinkel, ML Spano, WL Ditto, JN Weiss (1992) Controlling cardiac chaos. *Science* 257:1230-1235.

DP Lathrop, EJ Kostelich (1989) Characterization of an experimental strange attractor by periodic orbits. *Physical Review A* 40:4028-4031.

M Le Van Quyen, J Martinerie, C Adam, FJ Varela (1997) Unstable periodic orbits in human epileptic activity. *Physical Review E* 56:3401-3411.

K Narayanan, RB Govindan, MS Gopinathan (1998) Unstable periodic orbits in human cardiac rhythms. *Physical Review E* 57:4594-4603.

CA Skarda, WJ Freeman (1987) How brains make chaos in order to make sense of the world. *Behavioral and Brain Sciences* 10:161-195.

P So, JT Francis, TI Netoff, BK Gluckman, SJ Schiff (1998) Unstable orbits: a new language for neuronal dynamics. *Biophysical Journal* 74:2776-2785.

P So, E Ott, SJ Schiff, DT Kaplan, T Sauer, C Grebogi (1996) Detecting unstable periodic orbits in chaotic experimental data. *Physical Review Letters* 76:4705-4708.

Chapter 11

Other approaches based on the state space

In this chapter we continue our brief summary of assorted approaches based on the concept of state space. This material should be treated as an introduction and guide to the research literature, as opposed to most of the previous chapters which are largely self-contained presentations of the most widely used computational tools. In some cases the material presented here is very promising but has yet seen little application to physiology.

11.1 Properties of mappings

In section 9.5 we discussed, in the context of multiple time series, the ideas of continuity and differentiability in mappings (functions taking one time series or attractor to another). A general and fruitful idea based on continuity is that points that are close together should map to points that are also close together (that is, two state-space points in one system should be transformed by a smooth continuous function to two state-space points that are close together in the second system) (Pecora *et al.* 1995). The same concept can be applied to test for continuity, and by implication smoothness and determinism, in a single system. Again the idea is that if two points are found in the state space that are very close together, then these points should lead in the future to two points that are also close together: if $x(j)$ and $x(k)$ are two attractor points that are very close together, then $x(j+1)$ and $x(k+1)$ should also be close together, relative to the distribution of inter-point distances on the attractor. Part of the practical problem in implementing this idea comes from defining "closeness," and determining when points are closer together than would be expected by chance. The simple notion of continuity is also made difficult in practice by finite data sampling, so that the idealized limiting case of identifying points that are arbitrarily close together runs into problems.

Here we present one example of a computational tool that has been proposed to test for continuity. The method of exceptional events in section 11.3 below is another implementation of the same general idea.

A novel way to assess continuity was suggested by Wayland et al. (1993). This method begins by choosing a reference point $x_0(j)$ on the attractor, and finding a set of k nearest neighbors $x_i(j)$, $i=1...k$. The idea is that if all of these points are sufficiently close together, all of their projections one time step ahead should be of approximately the same size. (Most continuity tests are based on the assumption that these projected points should remain close together.) So, the projections of these points one step ahead, $x_i(j+1)$, are determined, and the lengths of these projections, or translations, is computed: for the ith neighbor $x_i(j)$ and its projection $x_i(j+1)$, the translation distance is $d_i=|x_i(j)-x_i(j+1)|$, where the vertical bars indicate Euclidean distance in the M-dimensional embedding space.

It is a simple matter to compute the average translation amplitude:

$$<d> = \frac{1}{k+1}\sum_{i=0}^{k} d_i$$

(where d_0 is the translation of the reference point itself), and the translation error:

$$e_{translation} = \frac{1}{k+1}\sum_{i=0}^{k} \frac{|d_i - <d>|^2}{|<d>|^2}.$$

The translation error represents the deviations of the translation amplitudes from the mean, normalized by the mean. For a deterministic system the translation amplitudes should all be approximately equal, and $e_{translation}$ will be small. This local measure is made into a global measure by finding $e_{translation}$ at several reference points randomly spaced on the attractor, and taking the median of these values.

11.2 Parallel flows in state space

In Chapter 10 the observation was made that, with a deterministic system, trajectories in the state space tend to form a coherent structure rather than a haphazard tangle. Another aspect of this property is that trajectory paths in any small part of the state space tend (on average) to be aligned with each other. This is obvious in the case of the Lorenz attractor, for example; in the right panel of Fig. 11.2.1, in any small volume of the state space the paths are approximately parallel to each other.

This property can be quantified by finding local "vector fields" that correspond to the average trajectory "flow" in a small volume (Kaplan & Glass 1992, 1993). This is done by dividing the state space into discrete regions ("coarse-graining") and finding the average trajectory direction in each region. To do this, a unit vector is established in each region every time that the trajectory passes through it; the length of the vector is one, and the direction is determined by the locations where the trajectory enters and exits the region. After a unit vector is assigned to each trajectory pass through that region, the average of these unit vectors is computed, and its length determined. If the multiple trajectory passes through a given region are highly aligned, the unit vectors will be nearly parallel to each other, and the resultant average vector in that region will have a length close to one. If the trajectory passes through a region are not aligned, then the resultant average vector will have a length much less than one. The lengths of these average vectors as a function of the number of times that the trajectory passes through a region provide a measure of the degree of alignment of the paths; as the number of trajectory passes increases, the resultant average length will decrease rapidly for random data and slowly or not at all for a deterministic system. Thus the statistics of the lengths of these average vectors, relative to a null hypothesis of random behavior, can give an indication of the degree to which the trajectory paths are aligned, reflecting in turn the deterministic nature of the underlying system. (This approach was later generalized by Salvino and Cawley (1994).)

A depiction of the concept can be found in Fig. 11.2.1. To the right is a two-dimensional time-delay reconstruction of the x variable of the

Lorenz system, with a 20x20 grid overlaid to create the coarse-grained regions. The length of the vector average in each of these regions is shown in the graph to the left. In regions where the trajectory paths are closely spaced and aligned – near the extremes of each lobe – the average vectors are large. Near the center of the attractor, where the trajectory paths go in both directions within close proximity, the average vectors are smaller. The same presentation for Gaussian random data is in Fig. 11.2.2, where the average vector lengths are shorter due to the haphazard arrangement of the trajectory paths as they near each other. (In these graphs, the data have been normalized to a range of 0-20 so that the coarse-grain regions can be formed from a grid of unit-size boxes in each case.)

While it provides an interesting example of intuitive reasoning applied to the state space, this technique has not seen wide use in physiology. One of its problems is that deterministic structure is often best revealed by those relatively rare occurrences of two points on the attractor, separated in time, being close together. The vector-averaging technique does not make good use of these point pairs, since they can easily fall into different regions when the state space is subdivided (DT Kaplan, personal communication).

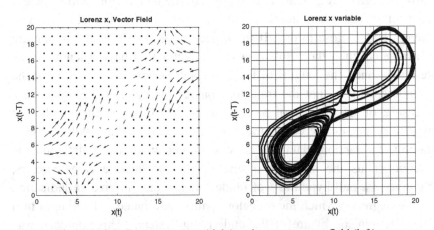

Figure 11.2.1. Lorenz attractor (right) and average vector field (left).

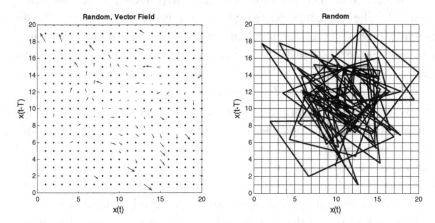

Figure 11.2.2. Random data "attractor" (right) and average vector field (left).

11.3 Exceptional events

An alternative approach to detecting determinism is the method of "exceptional events," or the "delta-epsilon" method (Kaplan 1994). This method rectifies one of the drawbacks of the vector-average method of section 11.2, which does not make effective use of nearby points which are important in providing information as to determinism. Here, points on the reconstructed attractor that are very close together are identified; these points are then each projected one time step into the future, and the new points examined to see how close together they are. (By projecting one time step into the future, we simply mean taking the next point in time along the same trajectory.) In the limiting case, as the initial points get closer together, the projected points should also get closer together, if the system is deterministic. (The name "exceptional events" comes from the fact that the occurrence of points that are extremely close together will be relatively rare, or exceptional, and it is the behavior of these points that is of interest.)

Alternatively, if one assumes that the data set is generated by a random process, a suitable random model can be created and several surrogate data sets generated from it. Points close together in the random data set are then each projected one time step ahead, as for the actual data set. The projected points may again get closer together as the original points get closer together, but only to a certain extent; in a random system the distance between the projected points will not converge to near zero in the limit, because of the presence of the significant (relative to any deterministic structure) random component.

By selecting appropriate stochastic models as null hypotheses (see Chapter 6), the likelihood that the original data set was generated by a system with deterministic dynamics can be approximated. Fig. 11.3.1 shows an analysis of a deterministic sine wave (top) and random noise (bottom). The deterministic case is distinguished by the fact that the curve drops toward zero at low values of initial distance (delta, δ), indicating that points that are close together (small δ) project to future points that are also close together (small epsilon, ε).

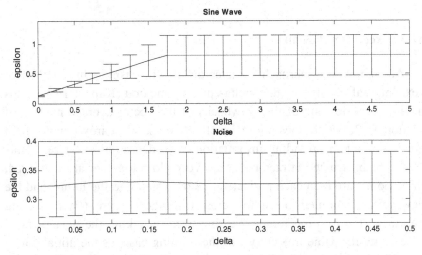

Figure 11.3.1. Example of Exceptional Events analysis of a random signal and a deterministic sine wave. For the deterministic system, as distance between two initial points (δ) decreases, so does distance of one-step ahead projections of those points (ε).

11.4 Lyapunov exponents

Informed readers may notice the conspicuous absence of Lyapunov exponents throughout this text. Lyapunov exponents quantify the rate of divergence of nearby trajectory paths in the state space, giving a measure of *sensitive dependence on initial conditions*, which is a hallmark of deterministic chaos. Algorithms exist to make these measurements (that in Wolf *et al.* 1985 is among the first). Some of the earliest papers that attempted to make use of Lyapunov exponents remarked on the difficulty of finding good estimates for the values of the exponents, and the sensitivity of the values to the computational parameters used to produce them. Although the methodology has improved and Lyapunov exponent values appear regularly in the research literature, it is the author's contention that this remains an area for more advanced investigators and so is outside the scope of this work.

One recent approach based loosely on the notion of Lyapunov exponents has been proposed as a test for determinism (Binder *et al.* 2005). This is based on the fact that, for a chaotic system, the distance between two trajectory paths that are initially a distance d_0 apart will increase with time as $d(t) \sim d_0 e^{\lambda t}$. This is the basis of the definition of the Lyapunov exponent, roughly equivalent to λ. For a chaotic system, the distance $d(t)$ should increase linearly with initial distance d_0, while for a random system $d(t)$ will be independent of d_0. Promising results were found with as few as 100 data points. There is an obvious similarity of this approach to the method of Exceptional Events above.

11.5 Deterministic versus stochastic (DVS) analysis

Another approach to assessing the relative amount of deterministic dynamics in a system is based on the ideas of nonlinear forecasting (Chapter 7). It has come to be known as *deterministic versus stochastic* (DVS) analysis (Casdagli 1992).

In this method, one-step ahead forecasting error is computed as a function of the number of nearest neighbors k used to form the forecast. As k increases, for a random system the forecasting improves, while for a chaotic system the forecasting becomes progressively worse. This is due to the fact that, with chaotic sensitive dependence on initial conditions, as more neighbors of the starting point are taken, more of these neighbors are farther from the starting point, and they will yield worse rather than better forecasts. For a random system, on the other hand, more neighbors (larger k) results in better statistics for the forecasting.

An example of the method is shown in Fig. 11.5.1, applied to data from the Lorenz system (top) and to random noise (bottom). As expected from a chaotic deterministic system (Lorenz), forecasting error increases as the number of neighbors increases. The opposite trend holds for the random data.

In practical applications, this procedure would be carried out over as wide a range of k values as possible, and repeated for different embedding dimensions.

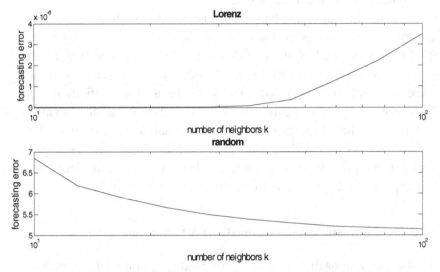

Figure 11.5.1. Example of the DVS test, applied to the Lorenz x variable (top) and to uncorrelated noise (bottom). In each case, the mean forecasting error is plotted as a function of the number of neighbors k used in the forecast.

References for Chapter 11

P-M Binder, R Igarashi, W Seymour, C Takeishi (2005) Determinism test for very short time series. *Physical Review E* 71:36219-36222.

M Casdagli (1992) Chaos and deterministic versus stochastic non-linear modeling. *Journal of the Royal Statistical Society B* 54:303-328.

DT Kaplan (1994) Exceptional events as evidence for determinism. *Physica D* 73:38-48.

DT Kaplan, L Glass (1992) Direct test for determinism in a time series. *Physical Review Letters* 68:427-430.

DT Kaplan, L Glass (1993) Coarse-grained embeddings of time series: random walks, Gaussian random processes, and deterministic chaos. *Physica D* 64:431-454.

LM Pecora, TL Carroll, JF Heagy (1995) Statistics for mathematical properties of maps between time series embeddings. *Physical Review E* 52:3420-3439.

LW Salvino, R Cawley (1994) Smoothness implies determinism: a method to detect it in time series. *Physical Review Letters* 73:1091-1094.

R Wayland, D Bromley, D Pickett, A Passamante (1993) Recognizing determinism in a time series. *Physical Review Letters* 70:580-582.

A Wolf, JB Swift, HL Swinney, JA Vastano (1985) Determining Lyapunov exponents from a time series. *Physica D* 16:285-317.

Chapter 12

Poincaré sections, fixed points, and control of chaotic systems

In this chapter we introduce the concepts of Poincaré section and fixed points. Although we have not made use of them so far in our exposition, these are important fundamental concepts. They are introduced here in the context of chaos control. The topic of chaos control is so vast, and is proving to be so useful, that we can only touch on the main points. This is one of the most active research areas in nonlinear dynamics, as the practical applications hold tremendous promise.

12.1 Poincaré section

The Poincaré section (or *surface of section*) is one of the more powerful devices for both qualitative and quantitative exploration of the dynamics of a system. In three dimensions, it is created simply by slicing the attractor with a two-dimensional plane, and indicating on the plane the locations where the attractor trajectory pierces it. (The extension to higher dimensions is straightforward, in each case the plane being one dimension less than the attractor.)

An example (that is unfortunately not very graphically interesting) is shown in Fig. 12.1.1. This is a Poincaré section from the Lorenz attractor. In the top panel the Lorenz attractor is shown, reconstructed with the time-delay technique. (Although the axes are labeled x, y, and z, these are in fact delay coordinates.) A plane is seen slicing the attractor at the level of $z=-12$. This plane is reproduced in the bottom panel, and the places where the attractor trajectory has pierced the plane (in the positive-going direction) are denoted by each plotted symbol (this creates a *Poincaré map*). Next to each symbol is the time order in which that point was created (the order in which the attractor crossed the plane at that point). Although the points lie almost on a straight line in this example (since the attractor is almost flat in this region), some interesting observations can be made about the dynamics based on the time order of the points. Note first that the first five points in time (labeled 1 to 5) move monotonically outward, toward larger values. This indicates that each time the trajectory returns to the bottom lobe of the attractor, it moves along a path that is farther away from an imaginary point near the center of that lobe. This expansion or stretching of the attractor is responsible for the sensitive dependence on initial conditions that is one hallmark of chaos. Also note that after the fifth point, they no longer progress monotonically along the plane, but instead move back and forth somewhat haphazardly. This shows that the attractor cannot expand forever but must from time to time be pulled back lest it expand to occupy all of the state space (this is what makes it an *attractor*).

It is important to observe that in creating the Poincaré section we have reduced by one the dimension of the system with which we are working. This can make further analysis easier, especially in a case like this one where the section is two-dimensional and easily comprehended. Of course the entire set of dynamics of the original system has not been retained in the reduced dimension of the Poincaré section, but in many cases it will capture important features and allow further analysis. This is especially the case for systems that are near-periodic, so that the plane crossings occur at near-constant intervals. In section 12.3 we will see how analysis of dynamics on the Poincaré section can provide a basis for control of a chaotic system.

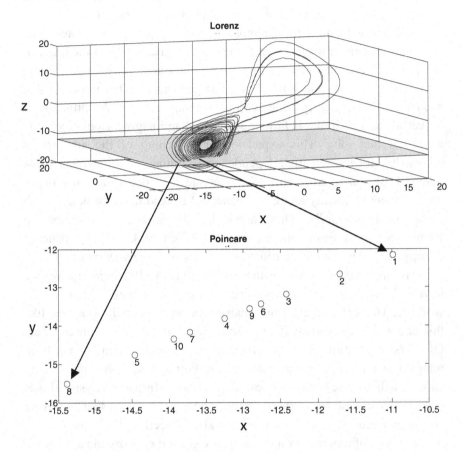

Figure 12.1.1. Example of a Poincaré section of the Lorenz attractor. Time-delay reconstruction of the attractor is shown (top), with intersecting plane at $z=-12$ shown in gray. This plane is reproduced in the bottom panel, showing the points where the trajectory has crossed the plane in the positive-going direction.

12.2 Fixed points

One of the principal ways to characterize a dynamic system is to identify its *fixed points*. As the name implies, a fixed point is a value (or a set of values making up a point in the state space) at which the system will stay, once having attained that point. (This is analogous to the notion of *eigenvectors* for a linear system. We will see this connection in the next section on control.) A fixed point can be *stable* or *unstable*. A stable fixed point will attract nearby points, while an unstable fixed point will repel nearby points. A simple one-dimensional example should help to make these concepts clear. In Chapter 1 we briefly saw the logistic equation, which is a one-dimensional mapping of the value at time i to the subsequent value at time $i+1$:

$$x(i+1) = \mu x(i)[1 - x(i)] .$$

This map is shown in the top panel of Fig. 12.2.1, for the parameter value $\mu=2$. Also on the graph is the identity line $x(i)=x(i+1)$; obviously the fixed point must be on this line. The fixed point indeed is at the intersection of the map and the identity line, at $x_F=0.5$. This value of the fixed point x_F can also be found analytically by setting $x(i)=x(i+1)$ in the defining equation above, and solving for x. This defines the fixed point since it is the value that does not change from time i to time $i+1$. (Fixed points for continuous-time systems are found by setting the derivatives dx/dt to zero, for the same reason, and solving for x (or for the multiple values that make up the state variables).)

We can find the stability of the fixed point by *linearizing* the function near the fixed point. In doing this we are assuming that the map can be described by a straight line near the fixed point. We find the slope of this line. If the slope is larger than 1.0 in magnitude, then points close to the fixed point will be thrown farther away from the fixed point on subsequent iterations, while if the slope is less than 1.0 the nearby points will move closer to the fixed point on subsequent iterations. ("Close" here means within the range where the linear approximation holds true.) Taking a little time to be sure that this is clear will be very useful in the long run.

So, let us take the derivative of the function that defines the progression of $x(i)$, that is, we find a linear approximation to the map that takes $x(i)$ to $x(i+1)$:

$$\frac{\partial x_{i+1}}{\partial x_i} = \mu(1 - 2x_n).$$

We then evaluate the derivative at the fixed point $x_F=0.5$, which results in a value of 0. Since this is less than 1 in magnitude, the fixed point is stable. This local linear approximation (slope) is shown as the tangent line to the map at the fixed point in the figure.

Now we carry out this procedure for a parameter value $\mu=4$. In this case, the map is shown in the next panel of Fig. 12.2.1. The fixed point is identified as before, $x_F=0.75$. However, this time the derivative evaluated at x_F is -2, which has magnitude greater than 1.0, and hence the fixed point is unstable at this parameter value. (The behavior of the values and stabilities of the fixed points, as μ varies, is a fascinating aspect of *bifurcation* analysis. The bifurcation diagram for the logistic map is almost as famous – and visually appealing – as the Lorenz attractor. Strogatz (1994) is the place to see this described simply and elegantly and in more generality.)

These stable and unstable behaviors are clearly seen in the bottom two panels in the figure, where the time series for $\mu=2$ and $\mu=4$ are plotted. Even though both series start at the initial point $x(1)=0.1$, for $\mu=2$ the function quickly reaches and stays at the stable fixed point, while for $\mu=4$ the fixed point is never actually attained (in fact the system is chaotic for this parameter value).

A slightly more complex example will help to solidify these points and provide the basis for the control example in the next section. At this point we need to delve into some more detailed mathematics than has been typical in this book. In Chapter 9 we saw the Hénon system, which is a two-dimensional discrete-time map: the values of x and y at time i are mapped to their future values at time $i+1$:

$$x(i+1) = A - x^2(i) + By(i)$$
$$y(i+1) = x(i).$$

Let us use the values $A=1.4$ and $B=0.3$. The resulting state-space map for the x and y components of this system in shown in Fig. 12.2.2.

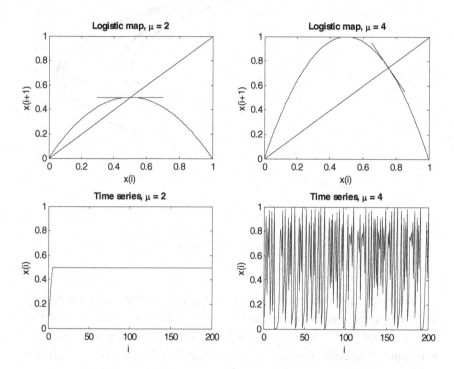

Figure 12.2.1. Examples of the logistic map $x(i+1)=\mu x(i)[1-x(i)]$, for parameter values $\mu=2$ (left) and $\mu=4$ (right). For $\mu=2$ the map has a stable fixed point at $x=0.5$. Stability is demonstrated by the slope of the map (zero) at the fixed point (top left); the time series $x(i)$ rapidly converges to the fixed point and stays there (bottom left). For $\mu=4$ the system is chaotic (right). The fixed point is unstable; the slope of the map has magnitude greater than 1.0 at the fixed point (top right). The time series (bottom right) exhibits chaotic behavior.

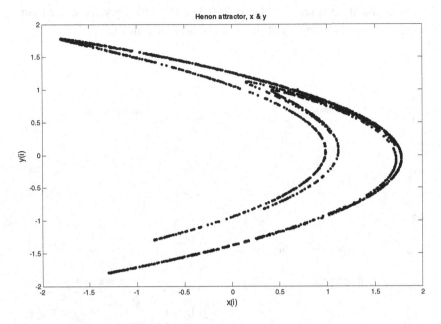

Figure 12.2.2. *x-y* map of the Hénon system.

The diagram shows the behavior of the two variables on successive iterations of the map. Note however that time information is lost in this plot: the point ($x(i+1), y(i+1)$) at time i could be far removed from the immediately preceding point ($x(i), y(i)$). That is why this is a *map* rather than a *trajectory* – consecutive points are not connected in a smooth curve as in the state-space trajectory of a continuous-time system.

Figure 12.2.3 is a *first-delay map* for the x variable from the same system. This shows how the value $x(i)$ is mapped to the subsequent value $x(i+1)$. It is analogous to the time-delay reconstruction of an attractor, with delay value set to one time step, but here as in the previous figure the points are distinct and should not be interpreted as a smooth trajectory in time. (There is an underlying smooth curve, which can be created by connecting the points. This curve is however still a *map* that defines subsequent iterated values, rather than a trajectory with respect to time. The curve is a fractal, as might be expected for this chaotic system.) This function appears to be multi-valued; that is, a particular value of $x(i)$ appears to map to several subsequent values $x(i+1)$. This is not actually the case – the y variable serves to resolve any ambiguity in the function, and information on y is missing from this graph.

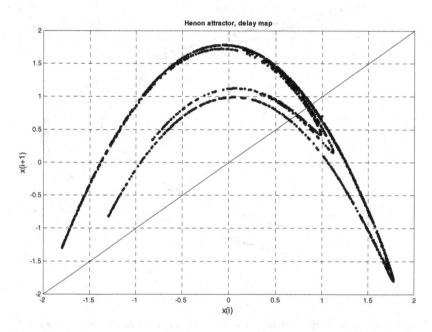

Figure 12.2.3. First-delay map of the Hénon system, showing how $x(i)$ maps to the next value $x(i+1)$. The identity line $x(i+1)=x(i)$ indicates where fixed points must lie.

As with the logistic map, the identity line $x(i+1)=x(i)$ is also shown. The fixed points must lie on this line. Setting $x(i+1)=x(i)$ in the defining equations for the Hénon system, and solving the resulting quadratic equation, yields $x_F=0.8839$ as the fixed point. (The other solution to the quadratic equation is $x_F=-1.58$, but the map does not visit this point.) This fixed point is one of the points of intersection of the map and the identity line in the figure. (Since the map is a fractal in this region, there are many other points of intersection of the identity line and the map. Not all of them are fixed points, however, because the map for the y variable must be at its fixed point at the same time as x is at one of these potential fixed points, and this is in general not the case.)

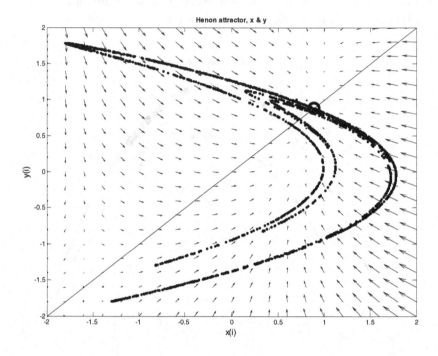

Figure 12.2.4. x-y map of the Hénon system, showing fixed point at (0.88, 0.88), and identity line $y(i)=x(i)$ which defines fixed points for this particular system (but not in general). The arrows show the "flow" of the variables at each point on the map; a point $(x(i),y(i))$ will move to subsequent point $(x(i+1),y(i+1))$ along the direction of the arrow at $(x(i),y(i))$.

Another way to visualize the effect of the fixed point is to plot the local discrete "flow" in the x-y plane, as in Fig. 12.2.4, along with the x-y map. Here, an arbitrary grid is created, with spacing of points in each direction at intervals of 0.2. At each of these (x,y) points, the values are plugged in to the Hénon equations, and the subsequent iterated point is found. The difference between the first and second points defines a local vector that shows in which direction, and by approximately how much (on a relative scale), the point (x,y) will move on the next iteration. These vectors are plotted for each starting point. This presentation is somewhat artificial because not all of the points on this grid are visited by the map. Nevertheless, it shows dramatically how the map acts to move points through the space toward the fixed point. This effect is very strong along an axis from approximately (2,-2) to (1,1), as indicated by the large arrows pointing almost directly to the fixed point (large circle) along this direction.

In order to characterize more fully the stability of this fixed point we must work in two dimensions. As before, we desire to find the derivative of the mapping function, to see if points near the fixed point are moved away from it or closer to it on subsequent iterations of the map. To do this we need a "two-dimensional derivative." The *Jacobian* is such an object. It is a matrix that shows the rate of change (derivative) of each state variable as a function of each of the other state variables. In two dimensions the Jacobian is:

$$\mathbf{J} = \begin{bmatrix} \dfrac{\partial x_{i+1}}{\partial x_i} & \dfrac{\partial x_{i+1}}{\partial y_i} \\ \dfrac{\partial y_{i+1}}{\partial x_i} & \dfrac{\partial y_{i+1}}{\partial y_i} \end{bmatrix}.$$

This is where the mathematics begin to get more involved, especially for those readers without previous experience with multi-dimensional linear systems. A complete understanding of stability and control (next section) requires familiarity with such concepts as eigenvectors and basis functions. The exposition to follow draws on these concepts but leaves out much detail. Hopefully all readers will be able to get a feel for the concepts sufficiently well to grasp the methodology of chaos control.

We now evaluate the Jacobian for the particular case of the Hénon map with $A=1.4$ and $B=0.3$, at the fixed point $x=0.8839$:

$$\mathbf{J} = \begin{bmatrix} -2x_i & B \\ 1 & 0 \end{bmatrix} = \begin{bmatrix} -1.7678 & 0.3 \\ 1 & 0 \end{bmatrix}.$$

In order to make use of this local derivative information in the space of the Poincaré section, we reduce this description to one with a more immediate geometric interpretation. This is done by finding the *eigenvalues* and *eigenvectors* of the Jacobian. For a matrix **A**, the eigenvalues λ and eigenvectors **e** are defined by this relation:

$$\mathbf{Ae} = \lambda \mathbf{e}.$$

This equation means that multiplication of an eigenvector **e** by the square matrix **A** (which produces another vector) reproduces the eigenvector, scaled by the factor λ. Let us interpret the vector **e** as describing a direction in space. Then, multiplication by **A** reproduces that same direction, with a magnitude altered by λ. Think of this operation in terms of the *x-y* space of the Hénon attractor (or on a Poincaré section). If the matrix **A** is in fact a Jacobian of some (in this case) two-dimensional function, then multiplication by **A** reflects the linearized function, defined by the Jacobian, operating on some point in space. Recall that we evaluate the Jacobian at the fixed point, and so this linearized approximation (multiplication by **A**) reflects the original function (map) operating on a point, and this approximation is valid near the fixed point.

The eigenvectors of the Jacobian matrix **J** determine "primary" or "special" directions in the *x-y* space (from the German "eigen," meaning "proper" or "inherent"). The eigenvectors define directions that are unaltered by iteration of the linearized function **J**. If a point lies along an eigenvector that originates at a fixed point of the map, then subsequent iteration of the function will move that point along that same vector, either toward or away from the fixed point (depending on whether the associated eigenvalue λ is greater than or less than 1.0 in magnitude). This will be true as long as the points are close enough to the fixed point for the linear approximation to remain valid. (These directions and operations are described in reference to the fixed point because the Jacobian was evaluated at the fixed point; the general features of eigen-directions are applicable throughout the *x-y* space but are of particular interest because of the importance of the fixed point.)

An important fact regarding eigenvectors *in this application* is that, for a two-dimensional system and a fixed point that is contained on the attractor, there will be an unstable and a stable eigenvector. Stability is determined by the eigenvalue associated with each eigenvector. If both eigen-directions were stable, the system would be attracted to the fixed point and that would be the extent of the attractor. If both eigendirections were unstable, the attractor would diverge and likewise not be very interesting or useful for the control aspects to follow.

The directions defined by the stable and unstable eigenvectors are usually described as the stable and unstable *manifolds* of the two-dimensional system. The traditional image is that of a saddle with the fixed point in the center. Along one direction the saddle slopes toward the center, so that a marble placed along that axis will move toward the fixed point. Along the perpendicular direction the saddle slopes away from the center, and a marble on that axis will move away from the fixed point at the center.

Now we see that the eigenvectors define the directions along which the system will continue, once placed there (analogous to a fixed point but in two dimensions, so a "fixed vector" in some sense). The associated eigenvalues tell us whether or not deviations from the fixed point along each eigenvector will be toward the fixed point (magnitude less than 1.0) or away from the fixed point (magnitude greater than 1.0).

The eigenvectors for the Hénon system fixed point are shown as thick lines overlaid on the x-y map in Fig. 12.2.5, and labeled as e_u and e_s for the unstable and stable directions, respectively. Also shown is the local vector flow as in Fig. 12.2.4 but only for points actually on the map.

It may seem counterintuitive, but the unstable eigenvector (associated with the eigenvalue with magnitude greater than 1.0) is along the direction with the largest arrows, where the (linearized) function acts the most strongly to pull points toward the fixed point. One might think that these large arrows, indicating strong flow toward the fixed point along that direction, would point along the stable eigenvector. However, the large arrows indicate instead that a point along this line will actually tend to overshoot the fixed point on the next iteration, rather than slowly converge to the fixed point. Movement along the stable eigenvector direction is more gradual. Fig. 12.2.3 shows this. For example, a value of $x(i)$ near 0 maps to subsequent values of 0.9 or greater (depending on the corresponding value of y), while a value near 1.5 maps to negative values, in each case overshooting the fixed point at 0.88.

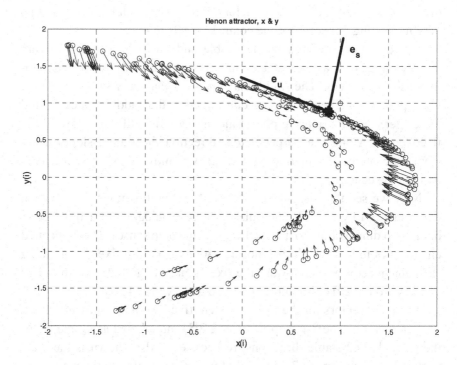

Figure 12.2.5. *x-y* map of the Hénon system, showing fixed point at (0.88, 0.88). As in Fig. 12.2.4, the arrows show the "flow" of the variables at each point, but only at those points actually visited by the map. The stable and unstable eigenvectors of the linearization at the fixed point, e_s and e_u, are also shown.

We end this section on fixed points and their stability by pointing out that a fixed point on a Poincaré section represents a periodic orbit of the larger attractor: each time the orbit crosses the Poincaré section it does so at the same location. The stability of the periodic orbit can be assessed by assessing the stability of the fixed point in the Poincaré section, as we have just shown. This observation is the basis for chaos control, which involves manipulating the system in order to place its state along a stable eigenvector and so lead it toward the fixed point, and periodic behavior.

12.3 Chaos control

As noted above, *chaos control* is one of the most active and promising lines of research stemming from the concepts of nonlinear dynamics and the state space. Control of a chaotic system is based on the observation (discussed in Chapter 10) that a chaotic attractor contains many (theoretically infinite) unstable periodic orbits (UPOs). These orbits, if they were stable, would create fixed points on a Poincaré section. If such a fixed point can be identified, and the behavior of the nearby points determined, then it may be possible to stabilize the unstable orbit (and fixed point) by making a small modification to some available system parameter. That is, when a point on the Poincaré section is close to the fixed point, we perturb the system to as to place the subsequent point at the fixed point, or along the stable manifold (eigenvector) that will lead to the fixed point.

Since a chaotic attractor is made up of many UPOs of different periodicities, we can pick one with desired properties and choose to stabilize it with this approach. We thereby make the system periodic. A subsequent small change in the system parameter can then be used to stabilize different orbits as desired. Therefore the presence of chaos allows great flexibility in choosing the desired performance as a consequence of these small control perturbations. This is not the case for non-chaotic systems, and there are obvious implications for how the brain may use neural chaos to perform some tasks requiring rapid selection from a large store of potential actions or perceptions.

The specific chaos control technique we describe here is the *OGY* method, named after the authors who first described it (Ott *et al.* 1990). The initial description (and the one we follow here) reflects a simplified version of the real world, because we assume that we have access to the defining equations of the chaotic system, and so can make all the necessary calculations analytically and ahead of time. In the real world, these quantities must be approximated from measured values before the control strategy can be computed. Nevertheless the concept and demonstration are significant steps in understanding and applying nonlinear dynamical concepts.

As in the previous section, this section is necessarily more mathematical than the rest of this book. Very nice summaries, in more detail and not as terse as the original paper, can be found in the wonderful book by Flake (1998), and in Spano and Ditto (1994).

We start by assuming that we have available to us a suitable Poincaré section (or other discrete-time map) from the system to be controlled. Here, we assume that such a map has been extracted from the system, and that the appropriate measurements can be made to characterize the dynamics on the map. In the specific derivation to follow, we pretend that the Poincaré section is described by a Hénon map (or that a discrete map such as this is what we wish to control), for which we also happen to have all the necessary measurements (derived from the defining equations). We also assume that the desired fixed point x_F (or $\mathbf{x}_F$ in vector notation) is zero; this simplifies the derivation without loss of generality.

Our goal is to derive a control strategy that places the system at the fixed point. We use the Jacobian, evaluated at the fixed point x_F, to approximate the behavior of points on the map near the fixed point:

$$[\mathbf{x}_{i+1} - \mathbf{x}'_F] \approx \mathbf{J}[\mathbf{x}_i - \mathbf{x}'_F]$$

where

$\mathbf{x}_i$ = point on Poincare map at time i

$\mathbf{x}_{i+1}$ = point on Poincare map at time i + 1

$\mathbf{x}'_F$ = new fixed point (after control perturbation)

$\mathbf{J}$ = Jacobian matrix.

We need to know how the fixed point will change due to small changes in the system parameter, which we call p in general (it is the Hénon parameter A in our particular case). The change in the fixed point as a function of the parameter p is given by:

$$\mathbf{g} = \frac{\partial \mathbf{x}_F}{\partial p}$$

so that (since $\mathbf{x}_F=0$):

$$\mathbf{x}'_F \approx \Delta p\, \mathbf{g}.$$

Inserting this expression into the first equation and rearranging:

$$\mathbf{x}_{i+1} \approx \Delta p\,\mathbf{g} + \mathbf{J}[\mathbf{x}_i - \Delta p\,\mathbf{g}].$$

In a completely non-obvious manner (see Flake 1998, Spano & Ditto 1994), we can expand the matrix-vector product on the right, to yield:

$$\mathbf{x}_{i+1} \approx \Delta p\,\mathbf{g} + [\lambda_s \mathbf{e}_s \mathbf{f}_s + \lambda_u \mathbf{e}_u \mathbf{f}_u][\mathbf{x}_i - \Delta p\,\mathbf{g}]$$

where λ_s, λ_u, and $\mathbf{e}_s$, $\mathbf{e}_u$ are the (stable and unstable) eigenvalues and eigenvectors of $\mathbf{J}$, and $\mathbf{f}_s$ and $\mathbf{f}_u$ are a set of *contravariant basis vectors*:

$$\mathbf{f}_s \cdot \mathbf{e}_u = \mathbf{f}_u \cdot \mathbf{e}_s = 0$$
$$\mathbf{f}_s \cdot \mathbf{e}_s = \mathbf{f}_u \cdot \mathbf{e}_u = 1.$$

For our purposes there are just a few things that we need to know about these new vectors $\mathbf{f}_s$ and $\mathbf{f}_u$. First, they allow us to express the multiplication in the equation for x_{i+1} above in a particularly useful way. Second, they are orthogonal to the original eigenvectors (first set of properties above). Finally, they are scaled appropriately to yield a dot product of 1.0 (second set of properties).

We know, through analytical derivation as here or through actual measurements, the values of the quantities in the equation above. In particular, we are going to use the control strategy to modify the parameter A of the Hénon system in order to stabilize the fixed point. We need to know the value of $\mathbf{g}$, the vector that tells us how the fixed point changes if we modify A:

$$\mathbf{g} = \frac{\partial \mathbf{x}_F}{\partial p} = \frac{\partial \mathbf{x}_F}{\partial A} = \begin{bmatrix} \dfrac{\partial x_F}{\partial A} \\ \dfrac{\partial y_F}{\partial A} \end{bmatrix} = \begin{bmatrix} 0.405 \\ 0.405 \end{bmatrix}.$$

Here now is the key to the control strategy. At time i we determine the point x_i where the trajectory crosses the Poincaré section. If this is suitably close to the desired fixed point x_F, we can calculate the required perturbation to be applied to the parameter A. ("Suitably close to the fixed point" is defined by the maximum allowable perturbation to A that we are willing to apply.) The perturbation is such that the subsequent point x_{i+1} is placed at the fixed point or *on the stable manifold* given by $\mathbf{e}_s$. In order to accomplish this, we want to place the point x_{i+1} so that its directed distance from the fixed point (that is, the vector from x_{i+1} to the fixed point) is orthogonal to the unstable direction given by $\mathbf{f}_u$, and this is done by making their dot product equal to zero:

$$\mathbf{f}_u \cdot \mathbf{x}_{i+1} = 0.$$

We make use of this new fact in the last equation above for x_{i+1}:

$$\mathbf{f}_u \cdot \mathbf{x}_{i+1} = 0 \approx \Delta p\, \mathbf{g} \cdot \mathbf{f}_u + \mathbf{f}_u \cdot [\lambda_s \mathbf{e}_s \mathbf{f}_s + \lambda_u \mathbf{e}_u \mathbf{f}_u][\mathbf{x}_i - \Delta p\, \mathbf{g}]$$

$$\Delta p\, \mathbf{g} \cdot \mathbf{f}_u = -[0 + \lambda_u \mathbf{f}_u][\mathbf{x}_i - \Delta p\, \mathbf{g}]$$

$$\Delta p\, \mathbf{g} \cdot \mathbf{f}_u [1 - \lambda_u] = -\lambda_u \mathbf{f}_u \cdot \mathbf{x}_i$$

$$\Delta p = \frac{\lambda_u}{\lambda_u - 1} \frac{\mathbf{x}_i \cdot \mathbf{f}_u}{\mathbf{g} \cdot \mathbf{f}_u}.$$

It is this final expression for Δp that is the crucial equation that defines the OGY control strategy. (Some details of vector-matrix manipulations have been skipped in this abbreviated derivation.) When $\mathbf{x}_i$ is close to the desired $\mathbf{x}_F$, the parameter p is adjusted by the amount Δp for one time step. This places $\mathbf{x}_{i+1}$ along the stable manifold leading to $\mathbf{x}_F$.

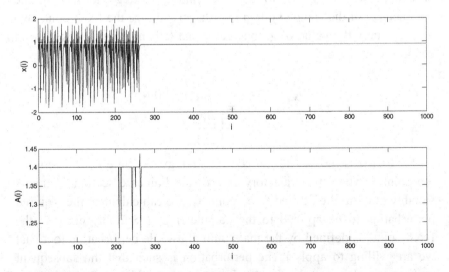

Figure 12.3.1. Example of OGY control applied to the Hénon system. Top plot is the x variable, bottom plot is the controlled system parameter A. Control starts at time $i=200$.

A demonstration of this control strategy in action can be seen in Fig. 12.3.1. The Hénon system is controlled so as to stabilize the fixed point at 0.88. The system is allowed to run for 200 iterations, and then the control law is applied. In fewer than 100 additional iterations, stabilization of the fixed point is evident in the top plot of the x variable. The bottom plot shows the history of the controlled parameter A, with a nominal value of 1.4. Just a few small perturbations to this parameter are sufficient to stabilize the fixed point, after which further control is unnecessary (in the absence of noise and on this time scale with high-resolution arithmetic computations).

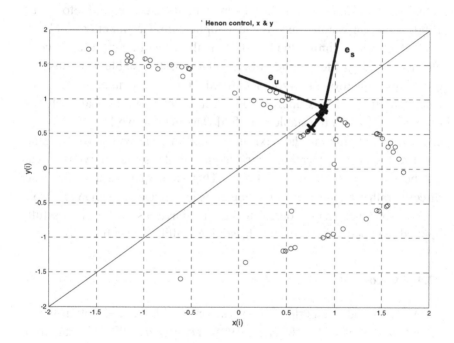

Figure 12.3.2. Example of OGY control applied to the Hénon system, showing action of the control strategy on the x-y map.

Figure 12.3.2 depicts the control action on the x-y map (acting as a Poincaré section in our demonstration). Those points marked by large **X**s depict the time course of one set of control perturbations, starting when an intial point is close enough to the desired fixed point (0.88, 0.88) for the control action to take place. As a result of perturbations to parameter A at each of the times corresponding to these points, the subsequent points move toward the fixed point, along a line that is orthogonal to the vector $\mathbf{f}_u$, which itself is orthogonal to $\mathbf{e}_s$ – in other words, along a line approximately collinear with stable eigenvector $\mathbf{e}_s$. The amplitudes of the perturbations are also appropriate to place the new points near the fixed point.

As we pointed out above, the field of chaos control is extremely active and several other control strategies have been devised. Control of actual mechanical systems has been demonstrated (e.g., Ditto *et al.* 1990). In a remarkable physiological application, Garfinkel *et al.* (1992) applied a strategy similar to OGY to a small piece of rabbit heart *in vitro*. On a first-delay plot of heart-beat intervals, the stable and unstable directions (manifolds) were identified. Given an interval ΔT_i, an electrical stimulus was applied to induce a "premature" beat such that $(\Delta T_i, \Delta T_{i+1})$ was on the stable manifold. The method works only in those cases where shortening the next anticipated interval will move the state point to the desired location, since obviously the same intervention can not be used to lengthen an interval. This technique perturbs the state directly, rather than a system parameter as in conventional OGY control.

We note that chaos control techniques have also been successfully applied to nonchaotic systems (Christini & Collins 1995a,b).

12.4 Anticontrol

A clever and counterintuitive variation on this general notion is that of *anticontrol* of chaos (In *et al.* 1995, Yang *et al.* 1995). Recall from Chapter 1 that one of the true insights revealed by nonlinear dynamics in physiology is that variability is normal and healthy, and a reduction in variability may be a sign of pathology. (The key of course comes in properly defining and quantifying *variability*.) Anticontrol takes advantage of this observation by applying chaos control techniques in order to keep the system on a chaotic attractor, rather than on a limit cycle or other pathologically low-variation trajectory.

One such anticontrol paradigm has been performed in rat hippocampal slices. This is one site of epileptic activity (Schiff *et al.* 1994), and hence an anticontrol strategy might prevent pathological entrainment and periodicity which could lead to a seizure. In this study, the authors were able to control the dynamics in a variety of ways with electrical pacing stimuli. Most notably, in a demonstration of anticontrol, they were able to keep the system *away from* a fixed point, and so avoid periodicities.

It remains to be seen to what extent these techniques will be more generally useful in complex biological situations where noise and high dimensionality impose significant complications.

References for Chapter 12

DJ Christini, JJ Collins (1995a) Controlling nonchaotic neuronal noise using chaos control techniques. *Physical Review Letters* 75:2782-2785.

DJ Christini, JJ Collins (1995b) Using noise and chaos control to control nonchaotic systems. *Physical Review E* 52:5806-5809.

WL Ditto, SN Rauseo, ML Spano (1990) Experimental control of chaos. *Physical Review Letters* 65:3211-3214.

GW Flake (1998) The Computational Beauty of Nature. Cambridge MA: MIT Press.

A Garfinkel, ML Spano, WL Ditto, JN Weiss (1992) Controlling cardiac chaos. *Science* 257:1230-1235.

VV In, SE Mahan, WL Ditto, ML Spano (1995) Experimental maintenance of chaos. *Physical Review Letters* 74:4420-4423.

T Kailath (1980) Linear Systems. Englewood Cliffs, NJ: Prentice-Hall.

E Ott, C Grebogi, JA Yorke (1990) Controlling chaos. *Physical Review Letters* 64:1196-1199.

SJ Schiff, K Jerger, DH Duong, T Chang, ML Spano, WL Ditto (1994) Controlling chaos in the brain. *Nature* 370:615-620.

ML Spano, WL Ditto (1994) The fundamentals of controlling chaos. *Integrative Physiological and Behavioral Science* 29:235-245.

S Strogatz (1994) Nonlinear Dynamics and Chaos. New York: Addison-Wesley.

W Yang, M Ding, AJ Mandell, E Ott (1995) Preserving chaos: control strategies to preserve complex dynamics with potential relevance to biological disorders. *Physical Review E* 51:102-110.

Chapter 13

Stochastic measures related to nonlinear dynamical concepts

Even more than the previous chapter, this one is a departure from the main theme of the book, which is nonlinear deterministic systems. Here we discuss some specific types of random (stochastic) systems. There are three main reasons for this digression. First, fractional Brownian motion (fBm) – a type of random process – can often mimic chaotic data. Second, many physiological systems have been found to exhibit fBm. Third, exploration of a system with the goal of finding deterministic dynamics sometimes leads to investigation of the random properties of that system as well.

13.1 Fractal time series, fractional Brownian motion

Just as a chaotic attractor forms a fractal object in state space, a time series itself can be a fractal. Here we consider the *self-similar* character of a fractal, in a statistical sense. This discussion will concentrate on a random fractal time series known as *fractional Brownian motion* (fBm), which is a generalization of conventional Brownian motion.

In 1828 botanist Robert Brown published his observations on the motion of microscopic particles within pollen grains. These motions were random, resembling the motion of dust particles suspended in air. Einstein, in one of his classic 1905 papers, provided some of the first mathematical descriptions of this *Brownian motion*, along the way establishing firmly the argument in favor of the atomic structure of matter: random bombardments of molecules could explain the properties of Brownian motion. (See review by Hänggi & Marchesoni 2005, and references therein.)

Mathematically, Brownian motion, or noise, can be thought of as the integral of Gaussian white noise (GWN). A *random walk* is such a motion in two dimensions. Fractional Brownian motion is a generalization, which can be created by the fractional integration of GWN. Needless to say, we will not delve into the mathematics of fractional integration and differentiation (Granger & Joyeux 1980).

fBm with a parameter H, which we term $B_H(t)$, has the following property: $B_H(at) \equiv a^H B(t)$, where $\equiv$ in this case means that the two processes have the same probability distribution: they are *statistically self-similar*. This defines the random fractal property of fBm. Another way to write this property is: $B_H(t) \equiv a^{-H} B(at)$, which says that if you adjust the time scale of the process by a factor a, then you must adjust the amplitude of the process by the factor a^{-H} in order to retain the original statistical properties, such as probability distribution.

The parameter H is known as the *Hurst exponent*; it can take on a value between zero and one. A value of $H=0.5$ produces conventional Brownian motion. Values greater than 0.5 yield a *persistent* process, which means that large values tend to follow large values, and small values tend to follow small values, on the average over many different time scales. On the other hand, a value less than 0.5 yields an *antipersistent* process, in which large and small values tend to alternate, again on the average over different time scales. Fig. 13.1.1 demonstrates this with some examples of fBm with different values of H.

It may seem odd, but the definition of fBm stipulates that $B_H(t)$ at any time t is Gaussian with zero mean. Therefore, the mean of an fBm process is also zero at all times. However, the standard deviation varies with time as a function of t^H (this can be seen in the figure, where the fluctuations of fBm increase with time). In a similar manner, the autocorrelation function $R_{xx}(t_1,t_2)$ is proportional to $[|t_1|^{2H}+|t_2|^{2H}-|t_1-t_2|^{2H}]$. That is, both the standard deviation and the variance depend on the specific time at which a measurement is made. In the case of the autocorrelation, the value depends not only on the difference between two times (as with stationary processes that we saw in Chapter 3), but also on the particular times t_1 and t_2. In other words, fBm is statistically *nonstationary*: its statistics change with time. However, the *increments* of such a process are stationary, and they have a Gaussian distribution (fractional Gaussian noise, FGN). Since the fBm process is not stationary, the power spectrum is hard to define in a rigorous manner. Nevertheless, a term $|2\pi f|^{-(2H+1)}$ appears in the frequency spectrum (actually the Wigner-Ville spectrum rather than the Fourier spectrum *per se*), and this leads to a power-law decay of the spectrum.

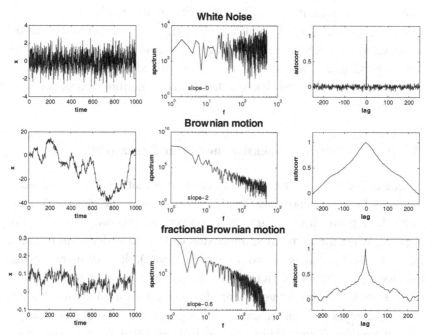

Figure 13.1.1. Examples of Gaussian white noise (GWN, top row), Brownian noise (center row), and fractional Brownian motion (fBm, bottom row). Sample time series are on the left, power spectra in the center column, and autocorrelations to the right.

As is the case with many deterministic measures, actual quantitative evaluation of the parameters of an fBm process can be challenging. In general terms these properties are often vastly simplified to the notion that both the autocorrelation and the power spectrum decay as a *power-*

law function. Instead of decaying with an exponential envelope ($\sim e^{-a\tau}$) as is common with many linear Gaussian systems, the autocorrelation decays as a power-law function of time ($\sim \tau^{-a}$). Likewise, the power spectrum decays as a power-law function of frequency ($\sim f^{-(2H+1)}$).

Signals such as fBm exhibit some strange properties. For example, there is the question of its very existence in a strict mathematical sense, since fBm does not have a proper derivative, and this implies that FGN (increments of fBm) can't exist. The derivative doesn't exist because, by the self-similar property, there are variations on all scales, no matter how small, and so continuity and differentiability are adversely affected.

It appears that, to a very high degree of fidelity, internet traffic can be described as a power-law process. There is an enormous literature on this topic (e.g., Cappé *et al.* 2002). It should be pointed out that 1/*f*-type processes are not necessarily generated solely by nonlinear systems (Voss 1978).

13.2 fBm, correlation dimension, nonlinear forecasting

One of the reasons that fBm is of concern is that it can mimic chaotic dynamics in some measurements. This is because some chaotic systems do produce power-law spectra, as do power-law random processes. For example, when the correlation dimension came into widespread use after the work of Grassberger and Procaccia (1983), it was more or less assumed that a random system, being of theoretically infinite dimension, will not yield a finite value of correlation dimension. This in fact was often used *de facto* as an indication that a given system was deterministic and not random. This bubble was burst when it was shown that certain types of random systems can yield finite values of correlation dimension (Osborne & Provenzale 1989): as embedding dimension is increased, the correlation dimension reaches an asymptote, as for a deterministic system, rather than increasing without bound.

The particular type of system that produces this anomalous scaling is one with a power-law spectrum, such as fBm. The reason that the dimension algorithm is "fooled" by this type of signal is that its state-space trajectory forms a random fractal curve, although not a fractal attractor. The correlation dimension does not depend on the order of the points in time, only on how they congregate in space, and so cannot distinguish between a random time series that produces a random fractal curve, and a deterministic time series that generates a fractal attractor. For the former the time ordering is random, but time order does not enter into the correlation dimension. We should note that this specific problem is not an artifact of the embedding method or limited amount of data; it is much more fundamental.

Theiler (1991) suggests some practical approaches to the issue of discriminating between chaos and power-law random processes. First is the observation that dimension measures should only properly be applied to a data set that is much longer than the "dominant period" of the system. For example, for the Lorenz system, one would expect to have data that covers both lobes of the attractor, otherwise the spatial properties are not sufficiently well represented. This is a problem for a system with strong power-law scaling, however, since there is effectively no dominant periodicity – power at low frequencies increases without bound (see the center column of Fig. 13.1.1). Thus, in principle, one cannot acquire enough data to adequately represent the system behavior.

Also, it is proposed that proper application of surrogate data techniques (Chapter 6) can help to distinguish between deterministic and random systems with power-law spectra. In fact, in the original study by Osborne and Provenzale, power-law scaling was analyzed by generating several random data series with specific power-law spectra – this is the generation of surrogate data, and the fact that the dimension estimates for these surrogates converge to finite values indicates that these estimates are not to be trusted (since it is known that they are random signals).

Another common analysis procedure, nonlinear forecasting, fares better in this regard. This was covered in section 7.6 of Chapter 7. To summarize, there is a distinct difference in the decay (with increasing forecasting step) of forecasting quality, depending on whether the system is chaotic or fBm.

13.3 Quantifying fBm: spectrum, autocorrelation, Hurst exponent, detrended fluctuation analysis

The two properties described in section 13.1 – power-law form of the spectrum and of the autocorrelation function – are often, in practice, used to define or at least to suggest that a given process is fractional Brownian motion. It is also common to measure the Hurst exponent by assessing the *rescaled range*: $R/S \propto T^H$, where R and S are the range and standard deviation of the integrated data in a window of length T (Addison 1997, Bassingthwaighte *et al.* 1994).

It can be shown that $H=(1+a)/2$, which expresses the relation between time-series scaling exponent H and frequency scaling exponent a (Rangarajan & Ding 2000). (This relationship holds strictly for fractional Gaussian noise, which is the sequence of increments of fBm.) In practice, to avoid false positive indications of such scaling behavior, which can arise from the use of either H or a alone, both values should be computed and compared for agreement with the expression above.

Another way to quantify this scaling is with *detrended fluctuation analysis* (DFA: Peng *et al.* 1995, 2000). The data series is integrated, and a straight line is fit via linear regression to the integrated data within a time window of a given size. The RMS variation about the fitted line gives the *fluctuations* $F(n)$ for window size n. The slope of $F(n)$ versus n, on a log-log scale, gives a scaling exponent that is akin to H. DFA removes local trends through the least-squares regression fit, and so is relatively immune to nonstationarity, which might yield spurious long-term correlations.

This methodology has been applied to heart-beat intervals (Peng *et al.* 1995, 2000) and stride intervals in human walking (Peng *et al.* 2000, Hausdorff *et al.* 1995). A general finding is that, analogous to dimension results in deterministic systems, there is a breakdown of normal healthy power-law scaling in pathology.

While these methods are simple and practical, and have been used to good effect in many published studies, it should be noted that the rigorous measurement and quantification of such statistically "troublesome" signals as fBm is a vast and mathematically sophisticated topic (Fischer & Akay 1996, Pilgram & Kaplan 1998, Cappé *et al.* 2002).

13.4 Self-organized criticality

A concept related to power-law scaling is *self-organized criticality* (SOC: Bak 1996, Bak *et al.* 1987, 1988). A system that exhibits SOC is *self-organizing* – the system is not directed by an external "teacher" to its final state, but arranges itself naturally. It is also *critical* – it is dynamically balanced between rigid inflexibility and randomness: there is structural stability in response to perturbations, but at the same time the ability to react quickly and appropriately to perturbations. It has been proposed that such self-organization "on the edge of chaos" can endow sensorimotor systems with the ability to respond rapidly to changes in environmental conditions (Skarda & Freeman 1987). This is an appealing conception, but it is still controversial.

Some of the defining characteristics of SOC are scale-invariance, self-similarity, and a $1/f$ spectral signature. These properties are related, since by *scale invariance* we mean that there is no preferred spatial or temporal scale – fluctuations occur on all scales. This of course can lead to self-similarity, of which a power-law spectrum is one sort.

SOC has been found in physical systems that can be modeled as the operation of many independent units, with local interactions that can extend globally. Supporting evidence for this view in the nervous system comes from a study of amplitude fluctuations in 10 and 20 Hz brain oscillations (Linkenkaer-Hansen *et al.* 2001); the fluctuations exhibit power-law scaling over several minutes. The authors conjecture that SOC can explain these properties, and that SOC may allow for the creation of "neural networks capable of quick reorganization during processing demands." Freeman has found similar extended spatiotemporal correlations in multi-electrode recordings of rabbit EEG (Freeman 2004).

In the specific case of the oculomotor system, at least one study has claimed evidence for SOC in the neural system that generates sequences of predictive eye movements (saccades) in response to periodically alternating visual targets (Shelhamer 2005). The distributed nature of the neural pathways that implement predictive saccades lends some credence to this claim (Gagnon *et al.* 2002, Gaymard *et al.* 1998). This is similar in nature to the proposition cited above, that other distributed brain structures might produce fractal scaling in the EEG. The different neural structures that implement predictive saccades may operate over different time courses (Ding *et al.* 2002) and so lead to the scale-invariance which is a feature of SOC.

References for Chapter 13

PS Addison (1997) Fractals and Chaos, an Illustrated Course. Philadelphia: Institute of Physics Publishing.
P Bak (1996) How Nature Works: The Science of Self-Organized Criticality. New York: Copernicus Books.
P Bak, C Tang, K Wiesenfeld (1987) Self-organized criticality: an explanation of the $1/f$ noise. *Physical Review Letters* 59:381-384.
P Bak, C Tang, K Wiesenfeld (1988) Self-organized criticality. *Physical Review A* 38:364-374.
JB Bassingthwaighte, LS Liebovitch, BJ West (1994) Fractal Physiology. New York: Oxford University Press.
EN Bruce (2001) Biomedical Signal Processing and Signal Modeling. New York: John Wiley & Sons.
O Cappé, E Moulines, J-C Pesquet, A Petropulu, X Yang (2002) Long-range dependence and heavy-tail modeling for teletraffic data. *IEEE Signal Processing Magazine* 19:14-27.
M Ding, Y Chen, JA Kelso (2002) Statistical analysis of timing errors. *Brain and Cognition* 48:98-106.
R Fischer, M Akay (1996) A comparison of analytical methods for the study of fractional Brownian motion. *Annals of Biomedical Engineering* 24:537-543.
WJ Freeman (2004) Origin, structure, and role of background EEG activity. Part 2. Analytic phase. *Clinical Neurophysiology* 115:2089-2107.
D Gagnon, GA O'Driscoll, M Petrides, GB Pike (2002) The effect of spatial and temporal information on saccades and neural activity in oculomotor structures. *Brain* 125:123-139.
B Gaymard, CJ Ploner, S Rivaud, AI Vermersch, C Pierrot-Deseilligny (1998) Cortical control of saccades. *Experimental Brain Research* 123:159-163.

CW Granger, R Joyeux (1980) An introduction to long memory time series models and fractional differencing. *Journal of Time Series Analysis* 1:15-29.

P Grassberger, I Procaccia (1983) Measuring the strangeness of strange attractors. *Physica D* 9:189-208.

P Hänggi, F Marchesoni (2005) Introduction: 100 years of Brownian motion. *Chaos* 15:026101-1:5.

JM Hausdorff, C-K Peng, Z Ladin, JY Wei, AL Goldberger (1995) Is walking a random walk? Evidence for long-range correlations in stride interval of human gait. *Journal of Applied Physiology* 78:349-358.

K Linkenkaer-Hansen, VV Nikouline, JM Palva, RJ Ilmoniemi (2001) Long-range temporal correlations and scaling behavior in human brain oscillations. *Journal of Neuroscience* 21:1370-1377.

SB Lowen, MC Teich (2005) Fractal-Based Point Processes. Hoboken NJ: John Wiley & Sons.

AR Osborne, A Provenzale (1989) Finite correlation dimension for stochastic systems with power-law spectra. *Physica D* 35:357-381.

C-K Peng, JM Hausdorff, AL Goldberger (2000) In: J Walleczek (ed) Self-Organized Biological Dynamics & Nonlinear Control. New York: Cambridge University Press. pp. 66-96.

C-K Peng, S Havlin, HE Stanley, AL Goldberger (1995) Quantification of scaling exponents and crossover phenomena in nonstationary heartbeat time series. *Chaos* 5:82-87.

B Pilgram, DT Kaplan (1998) A comparison of estimators for $1/f$ noise. *Physica D* 114:108-122.

G Rangarajan, M Ding (2000) Integrated approach to the assessment of long range correlation in time series data. *Physical Review E* 61:4991-5001.

M Shelhamer (2005) Sequences of predictive eye movements form a fractional Brownian series - implications for self-organized criticality in the oculomotor system. *Biological Cybernetics* 93:43-53.

CA Skarda, WJ Freeman (1987) How brains make chaos in order to make sense of the world. *Behavioral and Brain Sciences* 10:161-195.

J Theiler (1991) Some comments on the correlation dimension of $1/f^{\alpha}$ noise. *Physics Letters A* 155:480-493.

RF Voss (1978) Linearity of $1/f$ noise mechanisms. *Physical Review Letters* 40:913-916.

BJ West (1995) Fractal statistics in biology. *Physica D* 86:12-18.

Chapter 14

From measurements to models

The topic of this chapter is the holy grail of data analysis – the system identification problem. Given a set of data, and some measurements made on those data, can we come up with a mathematical model (the differential or difference equations) to describe the system? The brevity of this chapter should be an indication that this problem has not been solved, at least not in general. We must satisfy ourselves with a few broad comments.

14.1 The nature of the problem

The system identification problem has been addressed with great success for linear systems (e.g., Box *et al.* 1970, Ljung 1987). Although it is still difficult, restricting the possible models to those that are linear vastly simplifies the problem. Linear models are in the form of linear differential or difference equations, and once the input and output variables are identified, along with suitable state variables, system identification consists of finding the parameters of these equations. There are many ways to do this, and linear regression, which we discussed in Chapter 2, is one simple example. Even within the limited class of linear systems, however, there are still important distinctions, such as those in which the current output is a function only of past inputs (moving average, MA), those in which the current output is a function of past outputs (autoregressive, AR), or a combination of the two (ARMA). Of course the underlying issue of identifying the inputs, outputs, and state variables is still crucial and relies more on the investigator's knowledge and intuition than on any mathematical algorithms. Finally, the issue of noise presents its own set of difficulties. Although it is common to assume additive Gaussian noise that is independent from sample to sample (GWN), other forms are possible, and again a basic understanding of the system to be modeled is often as important as any mathematical tools that can be applied. The underlying theory of the overall process is greatly aided by the existence of the *Wold decomposition theorem*, which states that a random system (such as a linear system with noise) can be divided into a linear deterministic component and a random component that is a linear function of previous uncorrelated values of a random variable (Gershenfeld 1999).

The nonlinear identification problem is infinitely more complicated. The basic problem is that there is an infinite number of different classes of nonlinear systems – how to know if a system is best modeled by polynomials, logarithms, or other nonlinear functions? Although there has been progress in fitting parameters to a given system model once it has been chosen, and in determining the model order (number of parameters), the model selection problem remains critical.

14.2 Approaches to nonlinear system identification

One of the most widely-applied general approaches to nonlinear system modeling is the so-called Volterra-Wiener approach (Marmarelis & Marmarelis 1978, Rugh 1981), in which the system response is modeled by a series of integral equations of successively higher order (containing polynomial terms of the input signal). While this method has been very successful in many applications, it is a nonparametric technique that assumes little or no knowledge of the underlying system. (Recent attempts have been made to relate this approach to parametric models: Zhao & Marmarelis 1998.)

Another general approach of this type (in fact the Volterra-Wiener approach is a subset) is the use of basis functions to describe a time series (e.g., Crutchfield & McNamara 1987, Judd & Mees 1995). Again, it is the choice of basis functions that is a key, and that depends on the user's knowledge of the system under study.

The problem remains, then, of selecting the best model. An interesting approach to this issue has been proposed (Yadavalli et al. 1999), which uses a genetic algorithm. A genetic algorithm is an iterative optimization procedure motivated by biological reproduction. An initial problem solution is proposed, which is given by a sequence of parameters. In this case it is a sequence of constants, variables, and operators (addition, multiplication, sine, log, etc.), chosen and assembled randomly. A population of these candidate solutions is generated, all following the same structure but randomly generated. Each candidate solution is tested for how well it models the data. The best candidates are selected and used to create the next generation of solutions, by "mating" solutions in pairs: parameters of one solution are combined with parameters from another solution, repeatedly until a new population is formed. Random "mutation" is then applied to each new solution by randomly varying some of the parameters. This methodology has the promise of being able to explore an extremely wide variety and size of models, at the expense of considerable computational burden.

None of these methods make use of the state-space approaches that have been described in this book, but rather operate on the data series itself without reference to spatial properties (although time-delay embeddings are sometimes used in order to deal with unknown and possibly high dimensionalities). It remains unclear if doing so would present an advantage in solving the general nonlinear system identification problem.

14.3 A reasonable compromise

We see that to address fruitfully the system identification problem we would do well to reduce our ambitions to an acceptable level, at least in the context of the present work. Perhaps we can make use of the methods described in previous chapters to make some assessments of which behavior in a system might be random or deterministic, and its dimension, and proceed from there. This might not yield a model in the way that we would prefer, but it can tell us what system variables might be subject to modeling, and the approximate order of the model. At that point, physiological insight will be needed to propose a specific model form – knowledge of the system under study is still the best source of a model structure. Then, some of the techniques mentioned in this chapter might be used to find the actual equations that reproduce those properties that were found to exhibit significant nonlinear determinism.

We are not aware of any extensive studies using this hybrid approach. A simple example of how one might begin comes from an analysis of optokinetic nystagmus eye movements (this reflexive eye movement response was described in section 4.5 of Chapter 4, and will be discussed in detail in Chapter 15). Referring to Fig. 4.5.1A, observe that OKN is made up of alternating slow and fast phases. Sequences of different parameters can be extracted from OKN, as for example fast-phase intervals, fast-phase end points, fast-phase start points, and slow-phase durations. Nonlinear forecasting was carried out on each of these sequences (Shelhamer & Gross 1998). It was found that the only sequence with any potential deterministic component is the sequence of fast-phase end points. This has an interesting biological interpretation, and serves to confirm previous studies on OKN, as we will see in the next chapter. In terms of system identification, it tells us that we might fruitfully devote some effort to finding a deterministic model for fast-phase end points, with the other parameters modeled as random processes.

References for Chapter 14

M Akay (1994) Biomedical Signal Processing. San Diego: Academic Press.
GEP Box, GM Jenkins, GC Reinsel (1970) Time Series Analysis: Forecasting and Control. San Francisco: Holden-Day.
JP Crutchfield, BS McNamara (1987) Equations of motion from a data series. *Complex Systems* 1:417-452.
N Gershenfeld (1999) The Nature of Mathematical Modeling. New York: Cambridge University Press.
K Judd, A Mees (1995) On selecting models for nonlinear time series. *Physica D* 82:426-444.

L Ljung (1987) System Identification: Theory for the User. Englewood Cliffs, NJ: Prentice-Hall.

PZ Marmarelis, VZ Marmarelis (1978) Analysis of Physiological Systems: The White-Noise Approach. New York: Plenum Press.

WJ Rugh (1981) Nonlinear System Theory: The Volterra/Wiener Approach. Baltimore: The Johns Hopkins University Press.

M Shelhamer, CD Gross (1998) Prediction of the sequence of optokinetic nystagmus eye movements reveals deterministic structure in reflexive oculomotor behavior. *IEEE Transactions on Biomedical Engineering* 45:668-670.

VK Yadavalli, RK Dahule, SS Tambe, BD Kulkarni (1999) Obtaining functional form for chaotic time series evolution using genetic algorithm. *Chaos* 9:789-794.

X Zhao, VZ Marmarelis (1998) Nonlinear parametric models from Volterra kernels measurements. *Mathematical and Computer Modelling* 27:37-43.

Chapter 15

Case study – Oculomotor control

In this chapter we begin the presentation of recent research results on a few specific physiological systems. The presentation starts with a survey of the author's own work on one type of reflexive eye movement. This is not to imply that there are no other studies on the nonlinear dynamics of oculomotor control. In fact several recent studies (Clement *et al.* 2002a,b, Akman *et al.* 2005), have made explicit use of nonlinear dynamical concepts in modeling and analysis of specific aspects of eye-movement control. Another study more similar in character to those of the author is also discussed in this chapter.

15.1 Optokinetic nystagmus – dimension, surrogates, prediction

We present here the analysis of reflexive eye movement data. The eye movements under study are known as *optokinetic nystagmus*, or OKN. This reflexive motion occurs when a visual scene moves homogeneously over the subject's entire field of view. In the laboratory, this is accomplished by sitting the subject inside a large drum, which rotates about an earth-vertical axis. The inside of the drum is covered with a random or a striped pattern. The eyes reflexively attempt to track the pattern, generating nystagmus. OKN is a common clinical measure of vestibular and oculomotor function.

The OKN samples analyzed here were obtained from normal human subjects with a visual field velocity of 60 deg/sec. An example eye-position signal was shown in section 4.5 of Chapter 4, and its main features are described there as well. The most obvious feature is the near-periodic alternation of fast phases and slow phases. It is obvious that

OKN is not periodic. The slow phases are somewhat stereotyped, and we understand them as an attempt to stabilize vision by compensating for head and/or world motion. The logic in the timing of the fast phases and their interaction with the slow phases, however, is intriguing and has not been explained in a completely deterministic manner. Deciphering to what extent OKN is deterministic and nonlinear is an important step in the mathematical modeling of OKN.

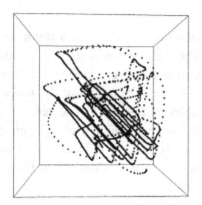 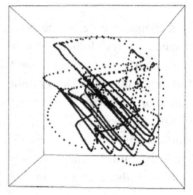

Fig. 15.1.1. Stereo view of OKN attractor, reconstructed from 4 seconds (2000 points) of data, with time-delay embedding. Time delay is 0.3 sec, sample rate 500 Hz.

A three-dimensional stereo view of the OKN attractor, reconstructed from 2000 points of the signal sampled at 500 Hz, is shown in Fig. 15.1.1. A vivid three-dimensional effect can be obtained by holding an index card between the two plots, resting one's nose on the card, and defocusing the eyes slightly as though looking at a distance. After a few seconds the plots should merge into a single view with a perception of depth. (A stereoscopic viewer may also be used.) Some of the features of this attractor were described in section 4.5, such as the high-density "sheet" of slow phases along the diagonal from upper left to lower right, and the fast phases that project out of this sheet with high velocity (points farther apart).

We present here an analysis of the OKN eye-position signal, as well as the analysis of various parameters extracted from OKN. These parameters are: slow-phase velocity, fast-phase amplitude, fast-phase interval, fast-phase starting position, and fast-phase ending position.

Recurrence analysis

A recurrence plot from 500 sample points (5 sec) of OKN data, sampled at 100 Hz, is shown in Fig. 15.1.2. The dominant feature of this plot is the large number of line segments that are almost parallel to the main diagonal. These line segments are caused by adjacent orbits on the trajectory that run close to one another for a period of time that is reflected in the lengths of the lines. These are due largely to the slow phases of the OKN signal; Fig. 15.1.1 shows that the slow phase trajectories are often parallel to one another for significant periods of time. The line segments are, however, in many cases longer than the typical time between fast phases (about 0.3 sec), and so the trajectories are parallel even across one or more fast phases.

A larger recurrence plot, from longer OKN recordings (Fig. 15.1.3), reveals other interesting features. There is large-scale structure in the form of small squares of high density, created from regions of the attractor in which the trajectory remains for a time, before moving on to a different region. A specific example might be the sheets of slow phases that are apparent in Fig. 15.1.1; presumably the high density of the slow phases creates such a pattern, while the fast phases take the trajectory from one such region to the other. Overall, there is also a decrease in density with increasing distance from the diagonal, in the larger recurrence plot, indicating that the long-term autocorrelations in OKN are weak and that OKN is not strictly periodic.

Recurrence matrices (recurrence plots without thresholding) for two of the parameter sequences are shown in Fig. 15.1.4. Each data set shows good stationarity (the plot density does not vary systematically as one moves away from the main diagonal), and little or no periodicity (no regularly-occurring diagonal dark bands). Occasional short diagonal line segments appear in the fast-phase end-point graph; this is indicative of a small amount of possibly deterministic structure in this data set.

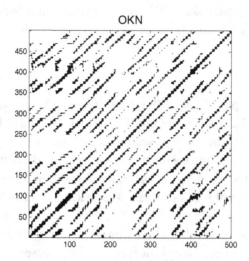

Figure 15.1.2. Recurrence plot of 500 points (5 sec) of OKN.

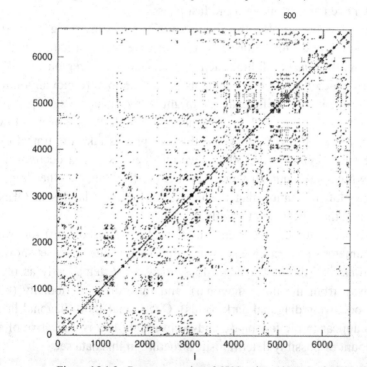

Figure 15.1.3. Recurrence plot of 6500 points (65 sec) of OKN.

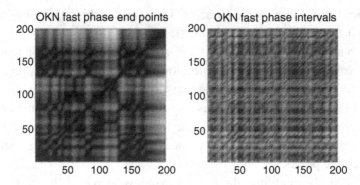

Figure 15.1.4. Recurrence matrices for OKN fast-phase end points and intervals.

Correlation dimension

An extensive analysis was performed to determine the computational parameters that provide the most reliable and consistent estimates of the correlation dimension. The values of several computational parameters were varied, to see their effects on the estimation of the correlation dimension: 1) sampling rate; 2) number of points; 3) time duration of the OKN segment; and 4) window length, $(M-1)L$, or the "embedding window," in the time-delay reconstruction. Obviously, these parameters are not independent.

Embedding windows of 0.3 sec and 0.6 sec were used. The first minimum of the autocorrelation function occurs at 0.18 sec; a window of 2 to 3 times this value is in the range 0.3 to 0.6 sec. The mean interval between fast phases – a kind of "pseudo-period" – is also on the order of 0.3 sec. As a check that the dimension estimates were robust with respect to changes in the length of the embedding window, additional windows ranging from 0.25 to 1.00 sec were tested on a portion of the OKN signal; dimension results were largely unaffected (less than ±10% variation).

A modification of the Grassberger-Procaccia algorithm, suggested by Theiler (1986), was also investigated. The modified algorithm does not consider points that are closer together in time than a certain threshold value when calculating the correlation integral. This avoids spurious correlations of points that happen to be temporally close rather than "dynamically" close on the attractor. The effect of this "Theiler window" (from 0.1 to 0.6 sec) was negligible (less than 10%).

At a sampling rate of 50 Hz, the correlation dimension has a tendency to increase with embedding dimension, indicating undersampling. This indicates the presence of a random component in the signal, which causes the attractor to expand as more dimensions become available in the embedding space. This random component is likely due to the undersampling, which destroys the deterministic relations between consecutive points.

At 100 Hz sampling, the 0.3 sec window provided the best overall results. We drew a 30-second sample with a dimension of 3.46 for further analysis.

Surrogate data

Several types of surrogate data were studied, based on the selected OKN sample; they are summarized in Table 15.1.1 and displayed in Fig. 15.1.5. Some of these surrogates serve as statistical controls on the dimension computations. They include: 1) **Shuffle**: created by shuffling the individual samples in the original OKN (to test the hypothesis that a random signal with the same amplitude distribution as OKN has the same dimension), 2) **Gaussian**: a Gaussian random process with the same mean and variance as the OKN (to test the hypothesis that a random process with identical second-order statistics has the same dimension), 3) **Phase-randomize**: formed by randomizing the phases of the frequency components in the spectrum of OKN (to test the hypothesis that a process with the same power spectrum or autocorrelation has the same dimension), 4) **AAFT**: adds the possibility of a static monotonic nonlinearity, and 5) **Pseudo-periodic**: assumes a noisy periodic structure for OKN. These surrogates give dimensions significantly removed from that of the original OKN data (Table 15.1.2). Therefore we reject the hypotheses given above, and conclude that OKN dimension values are not artifactual (i.e., not due to some random property of the data samples). Specifically, the third surrogate indicates that OKN is significantly different from linearly-correlated random noise (a random process with the same power spectrum as OKN). This is compatible with the hypothesis that OKN is a low-dimensional chaotic (nonlinear) system, but it is in no way a proof of this hypothesis.

Other surrogates tested specific hypotheses about the structure of OKN. They were created from the OKN data by shuffling fast and slow phases, to create three surrogate types: 1) **Fast-slow Shuffle**: segments containing a fast phase and the subsequent slow phase were randomly shuffled, 2) **Slow-fast shuffle**: segments containing a slow phase and the subsequent fast phase were randomly shuffled, 3) **Fast & slow shuffle**: fast and slow phases were shuffled independently. These surrogates are designed to test straightforward hypotheses about what in the structure of OKN determines its dimension (fast-slow coupling, slow-fast coupling, or the population of slow and fast phases). These three surrogates had dimensions *not* significantly different from that of the original OKN. Thus it is possible that OKN can be adequately modeled by one of the three processes that created these surrogates. Apparently, an essential property of OKN is the *population* of fast and slow phases, rather than any particular relationship between adjacent fast and slow phases.

The remaining surrogates consist of artificially generated signals that are meant to resemble OKN. The **Periodic** surrogate is simply a fast-slow combination, repeated several times to create a perfectly periodic waveform. The dimension algorithm did not yield a smooth scaling region with the attractor-reconstruction parameters used for the original OKN. This is because with successive periodic trajectories, the attractor points overlay each other completely, effectively producing an attractor with only as many points as it takes to complete a single cycle or orbit (about $N=300$). This results in substantial discretization of the correlation integral as criterion distance r increases. Resampling this surrogate at 500 Hz results in a dimension estimate near 1.0, as expected for a periodic signal. The **Random** surrogates are completely artificial and retain the second-order statistics (mean and standard deviation) of selected OKN parameters (see Table 15.1.1). The fact that the dimensions of these surrogates are less than that of the original OKN suggests that there are important dynamical properties that are not captured by these surrogates. In particular, temporal correlations in the fast-phase and slow-phase sequences are not retained in these surrogates, and this seems to be an important aspect of OKN (see below).

Name	Construction	Hypothesis
Shuffle	Shuffle the data points	OKN ≅ random process with same values
Gaussian	Surrogate has a Gaussian distribution with mean and variance of original	OKN ≅ Gaussian with same statistics
Phase-randomize	Surrogate phases have a Gaussian distribution with mean and variance of original	OKN ≅ process with same linear correlations
AAFT	Surrogate phases have a Gaussian distribution with mean and variance of original; amplitude ranks match original	OKN ≅ process with same linear correlations and static monotonic observation nonlinearity
Pseudo-periodic	Surrogate attractor orbits resemble original, fine structure is disrupted	OKN is periodic with uncorrelated noise
Fast-Slow shuffle	Scramble (fast-slow) segments	Slow-phase properties set by preceding fast phase
Slow-Fast shuffle	Scramble (slow-fast) segments	Fast-phase properties set by preceding slow phase
Fast & Slow shuffle	Scramble fast and slow segments independently	Population of fast and slow phases more important than their chronology
Periodic	Repeat a single (slow-fast) segment	OKN is periodic
Random 1	Artificial signal: fast-phase timing, amplitude, velocity, and slow-phase velocity have same mean and sd as original	Statistics of fast-phase timing, amplitude, velocity, and slow-phase velocity determine OKN
Random 2	Same as **Random 1** except that all slow-phase velocities identical	Statistics of fast-phase timing, amplitude, and velocity determine OKN

Table 15.1.1 OKN surrogate types.

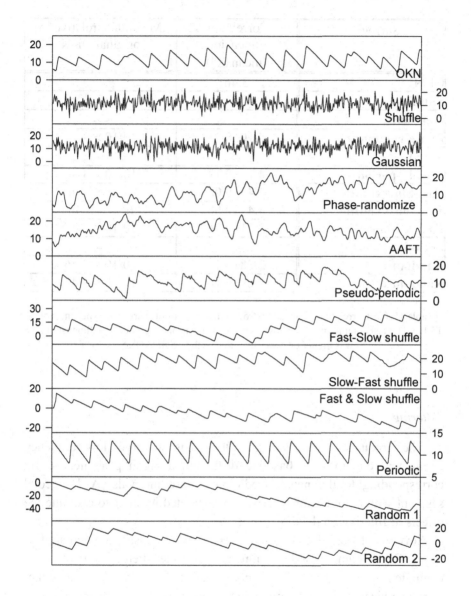

Fig. 15.1.5. One example of each of the OKN surrogates listed in Table 15.1.1.

Surrogate	Correlation dimension (mean±sd)	Magnitude relative to original OKN
OKN	3.46	
Shuffle	—	—
Gaussian	—	—
Phase-randomize	5.10±0.21	10/10 > 3.46
AAFT	—	—
Pseudo-periodic	4.15±0.13	10/10 > 3.46
Fast-Slow shuffle	3.43±0.08	3/10 > 3.46
Slow-Fast shuffle	3.47±0.06	6/10 > 3.46
Fast & Slow shuffle	3.31±0.06	10/10 < 3.46
Periodic	—	—
Random 1	2.88±0.11	10/10 < 3.46
Random 2	2.69±0.16	10/10 < 3.46

Table 15.1.2 Correlation dimensions from ten of each of the surrogate types in Table 15.1.1. The last column indicates how many of each surrogate produced a dimension that was higher or lower than that of the original OKN.

Filtering

The frequency spectrum of OKN has a large low-frequency component due to slow baseline drift, and a broad peak near 4 Hz corresponding to the near-periodic fast-phase intervals. A 10-second segment with a dimension of 3.07 was extracted in order to examine the effects of filtering on the dimension.

Several software digital low-pass filters were constructed and applied to this OKN segment, and the dimensions of the filtered segments were computed. Both recursive and nonrecursive filters were used, either applied twice in the standard manner, or applied once in the forward direction and once in the time-reverse direction in order to cancel phase changes induced by the first pass. The correlation dimension changes with the cutoff frequency of the filter, as shown in Fig. 15.1.6 and as predicted due to the fact that filtering adds a state variable to the system dynamics (see section 5.9).

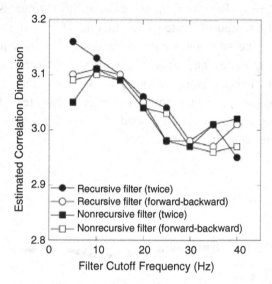

Figure 15.1.6. Effects of low-pass filtering on OKN dimension.

Nonlinear forecasting

Several data sets were extracted from a large segment of OKN, and future values of each sequence were forecast. The sequences are: slow-phase velocity, amplitude of each fast phase, time between fast phases, position of the eyes at the start of each fast phase, and position of the eyes at the end of each fast phase. Nonlinear forecasting was carried out as described in Chapter 7. There was little or no ability to forecast future values of the sequence of successive fast-phase intervals or sizes, or slow-phase velocities. There was some ability to forecast sequences of fast-phase starting positions and ending positions – there is some pattern to the eye positions at which fast phases begin and end. (These forecasting results are represented by bold lines in Fig. 15.1.7.)

Surrogates were generated from each of these sequences. These were phase-randomization surrogates, intended to help determine if the forecasting ability of these parameters reflected deterministic dynamics (in which case forecasting would be destroyed by the surrogates) or a correlated random process (in which case forecasting might be retained in the surrogates). Results from forecasting the surrogates (ten of each type) are represented as dotted lines in Fig. 15.1.7. It is apparent that forecasting of the surrogates and the original data are equally poor for fast-phase amplitudes and intervals, which demonstrates convincingly that these are uncorrelated random processes. On the other hand, the surrogates do not change the ability to perform short-term forecasting on fast-phase beginning and ending positions, which suggests strongly that these are correlated random processes.

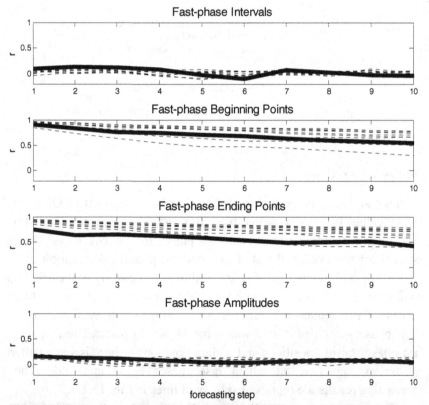

Figure 15.1.7. Nonlinear forecasting of sequences of OKN parameters. Forecasts were carried out from 1 to 10 steps ahead (abscissa). Correlation between actual and forecast value (r, ordinate) shows forecasting quality.

Mutual forecasting

Finally, an analysis was carried out to determine if there is dynamical coupling between any two of the OKN parameters. Since fast-phase beginning and ending points show some forecasting structure, emphasis was placed on these quantities. Mutual forecasting was carried out on pairs of OKN parameters. As described in section 9.3, in this procedure the points in one system (parameter sequence) are used to forecast the future course of a reference point in the other system. If this can be done, then the attractor of one system contains information about the other, and the two systems are dynamically coupled.

Results from mutual forecasting of fast-phase beginning and ending points are shown in Fig. 15.1.8. Here, instead of the correlation between actual and forecast values, the mean forecasting error is plotted, as a function of forecasting step. The upper left and lower right graphs show the results of forecasting *within* each data set, as in standard nonlinear forecasting. The upper right and lower left graphs show mutual forecasting: the ability of one data set to forecast the other. Surprisingly, the quality of the mutual forecasts is almost identical to that of the within-sample forecasts. This shows that these two OKN parameters are tightly coupled: one sequence provides significant information about the other. The identical forecasting procedure was carried out on phase-randomized surrogates of these two data sets. The results are shown in Fig. 15.1.9. Forecasting within each signal is unchanged, while mutual forecasting ability is completely destroyed (high error across all forecasting time steps). This is easy to understand when one realizes that the two parameter sequences were randomized independently, and then compared. Linear temporal correlations within each sequence are important, and they impart the ability to forecast future values; these correlations are retained in the surrogates. However, any *mutual* temporal structure – between the two sequences – is not retained in the surrogates. The beginning and ending positions are related to each other on a one-to-one basis and not merely in a distribution or population sense.

Nonlinear Dynamics in Physiology

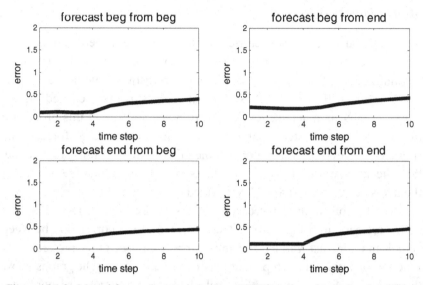

Figure 15.1.8. Mutual forecasting carried out on OKN fast-phase beginning and ending points. In each graph, the normalized mean forecasting error is plotted, as a function of the number of times steps into the future for which the forecast is carried out.

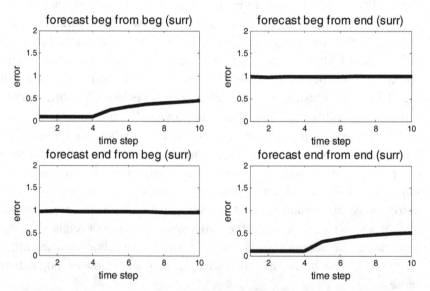

Figure 15.1.9. Mutual forecasting carried out on phase-randomized surrogates of the OKN parameters analyzed in Fig. 15.1.8.

Physiological interpretation

These analyses show that OKN eye movements may have some nonlinear and deterministic component, but also that randomness is a large part of the observed behavior. Surrogate analysis applied to the correlation dimension of OKN suggests that the population of slow and fast phases is more important than any specific relationship between adjacent slow and fast phases.

There was little or no ability to forecast the sequence of successive fast-phase intervals or sizes, or slow-phase velocities. There was some predictability in the sequence of fast-phase starting positions and ending positions – there is some pattern to the eye positions at which fast phases begin and end. This has some biological meaning, since the eyes make fast phases in order to pick up and track a portion of the visual field in which it is expected that something interesting will appear (Chun & Robinson 1978). This result lends some support to the idea that the fast phases of OKN are not programmed merely to interrupt the slow phases to keep them from exceeding the limits of eye rotation. Rather, the fast phases have an important role of their own – to choose an interesting and important point in the visual field at which to begin the next slow phase tracking. The essentially random (but correlated) structure of the end points suggests that a random strategy is well-suited for visual exploration, in the absence of better information. One way for the system to generate such a random-appearing sequence would be through the use of low-dimensional chaos, but it seems that this is not the case. However, it might be the case that the statistics of the end points attempt to match the statistics of contrast borders in the visual scene, in which case a more consistent and repetitive visual field (rather than the random rectangles used here) might change the end-point statistics.

15.2 Eye movements and reading ability

Although the literature on eye movements and reading is vast, there have been very few studies that have taken a nonlinear dynamics approach. In one such study (Schmeisser *et al.* 2002), the authors examined power spectrum, time-delay reconstructions, and correlation dimension of eye movement record obtained during reading.

Unfortunately, some of the classic shortcomings pointed out in previous chapters are present in this work, especially concerning the computation of the correlation dimension. Furthermore, as we have seen, these measures are not at all definitive in demonstrating the presence of deterministic chaos. Nevertheless, one interesting results does emerge, and that is that the dimension may be related to reading ability: the single dyslexic subject produced a low dimension (~2.1), while dimensions from the other subjects were larger. (The authors suggest that values from normal readers are in the range of 3.4-3.7, but examination of the results suggests instead that the dimension algorithm may not have converged to a finite value for these cases.)

The power spectra show a $1/f$ form, which as we have seen may be an indicator of chaotic dynamics, but might instead reflect randomness. It should be noted that other studies have found evidence for a $1/f$ type of power spectrum in oculomotor fixations during visual scanning (e.g., Aks *et al.* 2002).

Even with these limitations, the authors' overall conclusions are appealing and likely have some merit: "This suggests that the oculomotor behavior of subjects who read poorly is simplified and less free to vary or to adjust itself to task demands relative to the eye movements of normal readers. The loss of freedom to vary, demonstrated by this fractal analysis of eye movements, is quite analogous to that seen in epilepsy, cardiac dysrhythmia, and other disease states."

References for Chapter 15

OE Akman, DS Broomhead, RV Abadi, RA Clement (2005) Eye movement instabilities and nystagmus can be predicted by a nonlinear dynamics model of the saccadic system. *Journal of Mathematical Biology*.

DJ Aks, GJ Zelinsky, JC Sprott (2002) Memory across eye-movements: $1/f$ dynamic in visual search. *Nonlinear Dynamics, Psychology, and Life Sciences* 6:1-25.

K-S Chun, DA Robinson (1978) A model of quick phase generation in the vestibuloocular reflex. *Biological Cybernetics* 28:209-221.

RA Clement, RV Abadi, DS Broomhead, JP Whittle (2002a) A new framework for investigating both normal and abnormal eye movements. *Progress in Brain Research* 140:499-505.

RA Clement, JP Whittle, MR Muldoon, RV Abadi, DS Broomhead, O Akman (2002b) Characterisation of congenital nystagmus waveforms in terms of periodic orbits. *Vision Research* 42:2123-2130.

ET Schmeisser, JM McDonough, M Bond, PD Hislop, AD Epstein (2001) Fractal analysis of eye movements during reading. *Optometry and Vision Science* 78:805-814.

M Shelhamer (1992) Correlation dimension of optokinetic nystagmus as evidence of chaos in the oculomotor system. *IEEE Transactions on Biomedical Engineering* 39:1319-1321.

M Shelhamer (1997) On the correlation dimension of optokinetic nystagmus eye movements: computational parameters, filtering, nonstationarity, and surrogate data. *Biological Cybernetics* 76:237-250.

M Shelhamer (1998) Nonlinear dynamic systems evaluation of 'rhythmic' eye movements (optokinetic nystagmus). *Journal of Neuroscience Methods* 83:45-56.

M Shelhamer, N Azar (1997) Using measures of nonlinear dynamics to test a mathematical model of the oculomotor system. In: JM Bower (ed) Computational Neuroscience, Trends in Research, 1997, New York: Plenum Press. pp 833-838.

M Shelhamer, CD Gross (1998) Prediction of the sequence of optokinetic nystagmus eye movements reveals deterministic structure in reflexive oculomotor behavior. *IEEE Transactions on Biomedical Engineering* 45:668-670.

M Shelhamer, S Zalewski (2001) A new application for time-delay reconstruction: detection of fast-phase eye movements. *Physics Letters A* 291:349-354.

P Trillenberg, C Gross, M Shelhamer (2001) Random walks, random sequences, and nonlinear dynamics in human optokinetic nystagmus. *Journal of Applied Physiology* 91:1750-1759.

P Trillenberg, DS Zee, M Shelhamer (2002) On the distribution of fast phase intervals in OKN and VOR. *Biological Cybernetics* 87:68-78.

Chapter 16

Case study – Motor control

In this chapter we will discuss two aspects of human motor control, that is, the control of body and limb motion. The general outline of this material is drawn from an excellent review by Riley and Turvey (2002).

16.1 Postural center of pressure

Even when standing still, there are continuous variations in upright posture due to involuntary body sway. This sway can be measured by pressure sensors in a platform on which the subject stands, yielding a center-of-pressure (COP) signal (related to but not identical to center-of-gravity, since COP includes many other components of neuromuscular activity such as those arising from attempts to correct for large changes in center of gravity). Although the correlation dimension of COP profiles has been determined (Newell *et al.* 1993 found decreased dimension in the movement disorder tardive dyskinesia), we will concentrate here on other approaches to the dynamics of the postural control system.

Collins and De Luca (1993, 1994) investigated the COP profile from normal subjects as a "random walk," in which the mean-square displacement of COP as a function of time follows the same rule as for diffusion of a particle in space: $E(\Delta x^2) = 2D\Delta t$, where Δx is the change in COP along a given direction in the time interval Δt. The value of the *diffusion coefficient D* is easily determined by finding the slope of a graph of $E(\Delta x^2)$ vs. Δt, and dividing by 2. These graphs, for the case of static standing posture, exhibit two distinct scaling ranges or slopes, one at small time intervals (approximately one sec or less), and one at larger intervals. This suggests that there are different modes of control for short-term and long-term postural adjustments. Using techniques outlined in Chapter 13, the scaling exponent H was also determined for each of these ranges. In the short-term range, scaling exponents are large, indicating positively-correlated (persistent) variability. In the long-term range, smaller exponents indicate negatively-correlated (anti-persistent) variability, such that large deviations in one direction are corrected by motion in the other direction, as expected from closed-loop control. These scaling exponents also indicate that there are correlations over longer periods of time in this range, which might result from deterministic dynamics. This simple analysis suggests that there is a mixture of open-loop and closed-loop control over different time intervals, and possibly a mixture of random and deterministic dynamics as well. (However, see Peterka 2000 for an alternative interpretation.)

A later study by this group (Collins & De Luca 1995), as well as studies by others, showed that COP variability time series are less correlated (closer to white noise) with the eyes open than with the eyes closed. There is more instantaneous control and less reliance on past sway history with the eyes open. More extensive correlations over time with the eyes closed might reflect more determinism.

Further investigations of this phenomenon made use of recurrence quantification analysis (RQA, see Chapter 8). Riley and colleagues (1999) applied RQA to recurrence plots that were generated from reconstructed attractors, separately for COP data in the sideways direction (medio-lateral, ML) or the fore-aft direction (anterior-posterior, AP). The *percent determinism* measure (%DET) from RQA was between 0% and 100%, indicating that postural control is neither completely random nor completely deterministic. Postural variability (RMS sway) increased with the eyes closed, but so did %DET. In other words, with the eyes closed there is more sway but less of it is random. This may reflect a strategy in which the control system maintains tighter restraint on sway variation under this condition where vision cannot be used to aid stability, and so less random variation is allowed.

Balasubramaniam *et al.* (2000) followed up on this result by asking subjects to minimize their body sway in order to keep the projected spot of a laser pointer (attached to the body) within a target region while standing. The target was 1.1, 2.2, or 3.3 m distant, providing three levels of task difficulty (Index of Difficulty, ID). In Fig. 16.1.1 is the recurrence plot for one of these trials, showing patterned structure indicative of non-random dynamics. As seen in Fig. 16.1.2, performing the task decreased the amount of sway along the controlled direction (ML in this case, with the subject controlling the projected spot on the facing wall). Sway decreased more as the task got harder (increasing ID), and sway along the orthogonal AP direction increased as ML sway decreased. In Fig 16.1.3, decreased sway along the controlled direction is associated with a decrease in %DET, that is, a reduction in the amount of variability that can be attributed to deterministic dynamics. Although the effect is not large the trend is apparent. The need for increased precision in the task thus led to less sway but more randomness (relative to the total amount of sway). These effects reversed directions when the task was changed so that AP sway had to be controlled in order to accomplish the task. One interpretation of these results is that the deterministic component is reduced to a near-minimum value along the controlled direction, leaving only the random component and hence decreasing the %DET value.

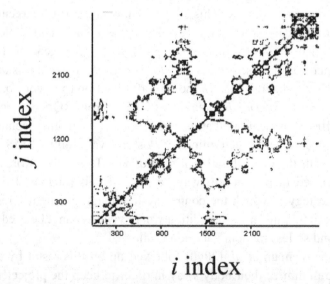

Fig. 16.1.1. Recurrence plot of sway variability (RMS) in the ML direction. (From Balasubramaniam *et al.* 2000.)

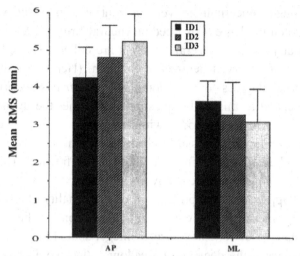

Fig. 16.1.2. Sway variability (RMS) in the AP and ML directions, while subjects attempted to control body sway in the ML direction with a laser-pointer task of three difficulty levels (ID). As task difficulty increased, sway along the controlled direction decreased (ML, right). (From Balasubramaniam *et al.* 2000.)

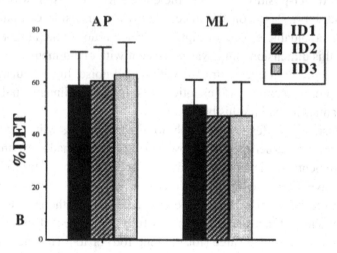

Fig. 16.1.3. Percent determinism (%DET), from recurrence quantification analysis (RQA), for the COP data from Fig. 16.1.2. As task difficulty increased and sway decreased, the percentage of the variability that could be attributed to deterministic dynamics increased (right, ML direction). (From Balasubramaniam *et al.* 2000.)

16.2 Rhythmic movements

A conceptual concern in movement control is how the very large number of potential degrees of freedom (neural firing patterns, multiple neuromuscular junctions, several muscles) are collectively reduced to a small number of observed degrees of freedom. (Here we will think of these observed degrees of freedom as the *dimensionality* of the behavior.) Related to this is the question of whether the dynamics of a rhythmic movement are periodic with additive noise, or if variability arises from deterministic (possibly chaotic) dynamics. Here we discuss some approaches to estimating the degrees of freedom (dimension) exhibited by a simple rhythmic behavior.

A natural approach to the problem of dimensionality of a behavior is, of course, to measure the correlation dimension of the associated attractor in state space. The next largest integer should tell us the number of state variables – the degrees of freedom – that would be needed to model that behavior. A study along these lines was performed by Kay (1988), who had subjects make repetitive paced single-finger movements. Time-delay reconstruction was used to generate the attractor from finger-position data, and the correlation dimension was found to be about 1.17. Based on the uncertainties in dimension computations and the unavoidable effects of high-dimensional noise, the author concluded that this dimension value was consistent with one-dimensional dynamics: a periodic limit-cycle attractor with added noise. In particular, there was no pronouncement of chaotic or other high-dimensional nonlinear dynamics to explain this behavior.

Using a different approach to the dynamics, Mitra *et al.* (1997) performed experiments in which subjects made rhythmic wrist movements while holding rods (36 or 66 cm long) that changed the effective limb inertia. The dimension of the resulting dynamics was assessed with False Nearest Neighbors (FNN) methods (see Chapter 4). FNN showed that the attractor was fully reconstructed in an embedding dimension of either three (for the long rod) or four (for the short rod). In other words, the movement behavior could be modeled with either three or four variables, depending on inertia. The relative lack of significant high-dimensional noise suggested that the dimension greater than 1.0

noted by Kay (1988) was in fact due to true dynamics, and not due to a noisy limit-cycle oscillator. (If the dynamics were a noisy limit cycle, then the number of false nearest neighbors would not drop to near-zero, as it does in this case.) Similarly, Goodman et al. (2000) showed a very small difference in required embedding dimension, using FNN, between wrist/rod movements at resonance and non-resonance.

These results were extended to the case of dynamical changes during learning of a behavioral task (Mitra et al. 1998). Subjects made simultaneous wrist motions while holding weighted rods to increase inertia. The task to be learned was making these motions 90 degrees (1/4 cycle) out of phase (between hands). There were three training and data-recording sessions over three days. As above, FNN was used to determine the number of active degrees of freedom – the dimension of the required embedding space. Although there was some change in FNN with learning, it was small.

A more convincing case was made by applying a modification of FNN: *local False Nearest Neighbors* (Abarbanel & Kennel 1993). The idea behind local FNN is that, in any *small region* of the attractor, the local dynamics may be adequately represented in a smaller space than is required to reconstruct the *entire* attractor. (It is this latter dimension – that required to reconstruct the entire attractor – that is determined by the conventional *global* FNN method. It is also the minimum acceptable embedding dimension M. The embedding dimension M is sufficient to properly reconstruct the entire attractor, while in any small region the dynamics may not require a dimension as large as M.)

To determine this smaller local dimension, d_L, first find the k nearest neighbors (the local region) of a reference point on the attractor, and project them into a smaller space d_L. (This process of treating the M-dimensional points as points in d_L-dimensional space can be thought of as simply dropping the appropriate number of components of the M-dimensional point, but there are some interesting complications that should be considered and are described in Abarbanel & Kennel (1993).) In this new d_L, use the original set of nearest neighbors to create a local forecast of the future course of the reference point (as in Chapter 7). Then, determine how accurate this forecast is, and label it a *bad prediction* (or forecast) if the error is greater than some threshold value (the threshold based on a percentage of the overall attractor size). Increase d_L and repeat this procedure, and also repeat for different values of k, until the proportion of bad predictions reaches an asymptote with increasing d_L and k. The value of d_L at which the asymptote occurs is the dimension that allows a proper reconstruction of the local dynamics.

In Fig. 16.2.1 are data acquired at various times during the learning of the wrist-coordination task. Moving from early to late (panels A to D, in order), we see that first the right hand achieves a stable rhythm, followed more gradually by the left hand. Local FNN analysis on these same data are in Fig. 16.2.2. In all cases, as the local dimension increases, the proportion of bad predictions decreases. There is a large additive noise component in the task variability during early learning (manifest by the fact that there continues to be a large proportion of bad predictions even at high local dimensions, where noise dominates). This dissipates as the task is fully learned (panel D), when deterministic dynamics are dominant. There is also a reduction in the effective number of degrees of freedom (d_L where asymptote occurs), as determined from the inflection point in each curve (the inflection indicates that any deterministic dynamics have been fully unfolded, and errors at higher d_L are largely the result of random noise). In particular, from panel C to panel D there is a change in inflection point from $d_L=5$ to $d_L=4$ for the right hand, while remaining at 5 for the left hand, showing that not only is the noise decreasing with learning, but the dynamics are decreasing their dimensionality. This novel approach to examining the dimensionality of the attractor avoids some of the problems in computing the correlation dimension.

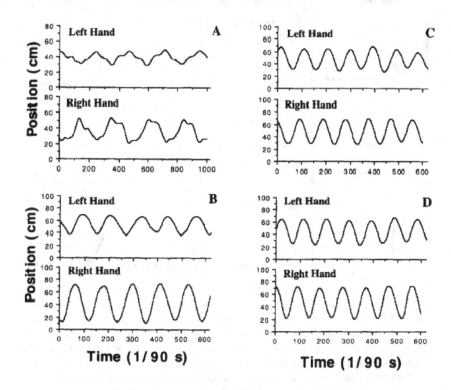

Figure 16.2.1. Motions of the right and left hands during learning of a coordination task (rhythmic motions 90 deg out of phase). Data samples are shown from four separate experiment sessions (chronologically A to D), showing the acquisition of learning. Movements become smoother and less variable as learning progresses. (Reprinted from *Human Movement Science*, vol. 17, Mitra, Amazeen, Turvey, Intermediate motor learning as decreasing active (dynamical) degrees of freedom, pp 17-65, Copyright 1998, with permission from Elsevier.)

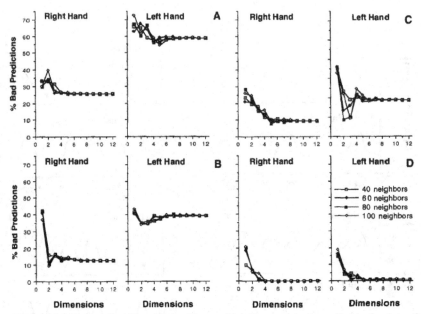

Figure 16.2.2. Results of local False Nearest Neighbors analysis applied to the data of Fig. 16.2.1. As learning progresses, the proportion of bad predictions at high dimensions decreases, indicating the decreasing effect of random noise. Once the task is fully learned (D), the dynamics can be reconstructed in a space with dimension of either 4 (right hand) or 5 (left hand). (Reprinted from *Human Movement Science*, vol. 17, Mitra, Amazeen, Turvey, Intermediate motor learning as decreasing active (dynamical) degrees of freedom, pp 17-65, Copyright 1998, with permission from Elsevier.)

References for Chapter 16

HDI Abarbanel, MB Kennel (1993) Local false nearest neighbors and dynamical dimensions from observed chaotic data. *Physical Review E* 47:3057-3068.

R Balasubramaniam, MA Riley, MT Turvey (2000) Specificity of postural sway to the demands of a precision task. *Gait and Posture* 11:12-24.

JJ Collins, CJ De Luca (1993) Open-loop and closed-loop control of posture: a random-walk analysis of center-of-pressure trajectories. *Experimental Brain Research* 95:308-318.

JJ Collins, CJ De Luca (1994) Random walking during quiet standing. *Physical Review Letters* 73:764-767.

JJ Collins, CJ De Luca (1995) The effects of visual input on open-loop and closed-loop postural control mechanisms. *Experimental Brain Research* 103:151-163.

BA Kay (1988) The dimensionality of movement trajectories and the degrees of freedom problem: a tutorial. *Human Movement Science* 7:343-364.

S Mitra, MA Riley, MT Turvey (1997) Chaos in human rhythmic movement. *Journal of Motor Behavior* 29:195-198.

S Mitra, PG Amazeen, MT Turvey (1998) Intermediate motor learning as decreasing active (dynamical) degrees of freedom. *Human Movement Science* 17:17-65.

KM Newell, REA van Emmerik, D Lee, RL Sprague (1993) On postural stability and variability. *Gait and Posture* 4:225-230.

RJ Peterka (2000) Postural control model interpretation of stabilogram diffusion analysis. *Biological Cybernetics* 82:335-343.

MA Riley, R Balasubramaniam, MT Turvey (1999) Recurrence quantification analysis of postural fluctuations. *Gait and Posture* 9:65-78.

MA Riley, MT Turvey (2002) Variability and determinism in motor behavior. *Journal of Motor Behavior* 34:99-125.

Chapter 17

Case study – Neurological tremor

This chapter provides an instructive example on the search for nonlinear dynamics in human hand tremor. Although there is an extensive literature on mathematical approaches to tremor, we restrict ourselves here to a small set of studies that applied the techniques in this book. The initial studies found preliminary evidence for chaotic dynamics in pathological tremor, while later studies with more careful computational procedures (by the same investigators) reversed this result. This is not the only case in which initial promising claims had to be tempered as procedures improved, but it is one of the more clear-cut.

17.1 Physiology background

Tremor is an involuntary rhythmic movement of a limb. There are several types. *Physiologic* tremor exists normally as a small, usually unnoticed, background tremor in normal subjects. *Essential* tremor might be thought of as an amplified form of physiologic tremor; easily visible, it develops slowly and is manifest in most cases as arm tremor of 4-12 Hz. ("Essential" refers to the fact that it has no apparent cause.) *Parkinsonian* tremor is one symptom of Parkinson's disease, a progressive movement disorder caused by dopamine deficiency. It is typically in the range of 4-6 Hz. Its early diagnosis is important because beneficial treatments are available.

The exact causes of these tremors are unknown. However, given the many feedback pathways in the body, and the time delays engendered by synaptic transmission times and axonal propagation delays, one might consider it surprising that such unwanted oscillations are not more common. (Feedback systems, with high enough gains, are a sure way to generate oscillatory behavior.) Essential tremor may arise from abnormal functioning of, or transmission from, a "central oscillator" in the inferior olivary nucleus, and almost certainly altered function of motor pathways through the cerebellum, brainstem, thalamus, and cortex are involved (e.g., Pinto *et al.* 2003). Problems in this same neural pathway might also explain Parkinsonian tremor, which likely also involves degeneration or abnormal functioning of the basal ganglia (Bergman & Deuschl 2002).

Characterization of the different types of tremor is important in order to gain insight into their neural mechanisms and, perhaps more importantly, for possible diagnostic purposes. As we will see, in particular, discriminating between essential tremor and Parkinsonian tremor has some difficulties, yet this ability would be very useful clinically. The nearly rhythmic nature of the different tremors lends itself well to power spectrum analysis, but other approaches have been used as well to gain more insight and discriminative power.

17.2 Initial studies – evidence for chaos

Examples of physiologic, essential, and Parkinsonian tremor are shown in Fig. 17.2.1, and formed the data set for an initial study into the dynamics of hand tremor (Gantert *et al.* 1992). Movements of the outstretched hand (loaded and unloaded) were measured with an accelerometer, sampled at 300 Hz for 35 seconds consecutively ($N=10240$ points in a data set).

Several lines of evidence showed that physiologic tremor has random and linear dynamics; that is, the signal can be represented as linearly filtered Gaussian white noise. A finite correlation dimension could not be found, suggesting a random signal, and a linear autoregressive model fit the data and could be used to forecast future values with high accuracy, suggesting linearity. Thus, one might interpret normal tremor as resulting from a damped linear oscillator (due to limb dynamics) driven by additive uncorrelated noise (due to random motoneuron firing).

The correlation dimension for Parkinsonian tremor was finite, ranging from 1.4 to 1.6, suggesting deterministic nonlinear dynamics and possibly chaos. The slopes of the correlation integrals for Parkinsonian tremor are shown in Fig. 17.2.2 (left) for M=2-5. With a main spectral peak below 10 Hz, and a sampling frequency of 300 Hz, there may be some concern as to the robustness of the results. In particular, the possibility of oversampling exists, which can introduce spurious correlations at small distance scales (e.g., Shelhamer 1997); this is the range (small r) over which power-law scaling exists in the figure. In addition, the time delay L used in the attractor reconstruction was 1/150 sec, which seems rather small given the sample interval of 1/300 sec, and no justification was given for this choice. As seen in Chapter 5, there are recommended procedures for choosing this parameter.

For essential tremor, the correlation dimension did not converge, as seen by the slopes of the correlation integrals given in Fig. 17.2.2 (right). This again suggests the presence of randomness. Also, the bispectrum of this signal was not zero, which is consistent with the dynamics coming from a nonlinear system. (The bispectrum (Bruce 2001) is a measure of correlation between two different frequency components. It should be zero for a linear system driven by white noise, since such a system will not create interactions between frequencies. (See Chapter 3.) The possibility remains that the system could be linear and driven by other than uncorrelated noise. Thus, with some exceptions, if the bispectrum is not zero for all frequency pairs, then the system is possibly not linear.)

Using a different set of measures, the same research group (Timmer *et al.* 1993, Deuschl *et al.* 1995) confirmed that the distinction between linear and nonlinear dynamics could be used to categorize normal and pathological tremors, respectively, while assessing different aspects of nonlinear dynamics was needed to separate different classes of pathological tremor (essential and Parkinsonian).

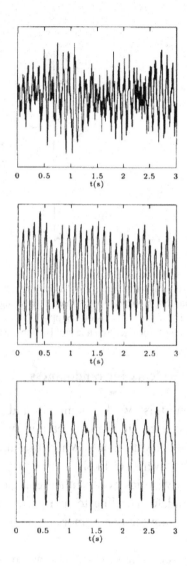

Figure 17.2.1. Examples of physiological tremor (top), essential tremor (center), and Parkinsonian tremor (bottom). (From: Timmer, Gantert, Deuschl, Honerkamp. Characteristics of hand tremor time series. Biological Cybernetics. vol. 70, 1993, Fig. 1, p 76, with kind permission of Springer Science and Business Media.)

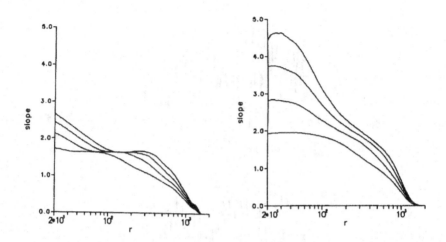

Figure 17.2.2. Slopes of correlation integrals for Parkinsonian tremor (left) and essential tremor (right). (From: Gantert, Honerkamp, Timmer. Analyzing the dynamics of hand tremor time series. Biological Cybernetics. Vol. 66, pp 483, 484, Figs. 4 and 6, with kind permission of Springer Science and Business Media.)

17.3 Later studies – evidence for randomness

Following up on this work, Timmer and colleagues (2000) endeavored to characterize in a more careful way the pathological tremors (essential and Parkinsonian). While the initial study described above showed that they were not likely to be noisy *linear* oscillators, it was unclear if their dynamics were either chaotic or produced by a random nonlinear oscillator (leading to non-periodic oscillations). Also unclear was whether or not the two tremors could be distinguished based on dynamical measures.

Data for this study consisted of hand tremor recordings, sampled at 1000 Hz for 30 sec (N=30000 points). The time delay L in the attractor reconstruction was selected as the first zero of the autocorrelation function, giving a value of about 50 time steps or 0.05 sec (compare this to 1/150 or 0.007 sec for the previous study discussed above). Reconstructed attractors can be seen in the left column of Fig. 17.3.1.

Power spectra of both tremor types exhibited broad peaks and harmonics, indicative of nonlinear oscillations with random variability (temporal variations in the periods of the oscillations broaden the spectral peaks, and harmonics indicate non-sinusoidal and hence non-linear oscillations).

To provide a basis for interpreting the measures of nonlinear dynamics, simulated data sets from two model systems were also examined. The first was the Lorenz system, which we have seen is chaotic with a dimension slightly greater than two. The second is the *van der Pol oscillator*, which is a nonlinear oscillator, to which has been added process noise (i.e., the noise enters into the dynamics and is not simply tacked on at the end as measurement error). These were meant to be representative of two possible explanations for pathological tremor – chaos versus a nonlinear random oscillator – and helped serve as a check on the discriminative power of the suite of computations.

Attempts to measure the correlation dimension of each tremor type were unsuccessful, as the algorithm did not converge but rather the slopes of the correlation integrals increased as the embedding dimension increased (Fig. 17.3.1, right column, panels c and d). This was true as well for the noisy van der Pol oscillator (panel b), while the Lorenz computations converged for sufficiently high embedding dimension (panel a). Thus neither type of tremor, in this improved analysis (relative to the initial study) seems to exhibit low-dimensional chaos. Likewise, the Poincaré sections and first-return maps from the tremor attractors did not show any of the well-defined structures that would indicate low-dimensional deterministic dynamics.

The data were also assessed with the method of Exceptional Events, or the δ-ε method (see section 11.3 in Chapter 11). Recall that this proceeds by finding those pairs of attractor points within distance δ of each other, then finding the distance ε as these points are projected ahead in time by one step. For a deterministic system with low noise, ε should go to zero as δ decreases, indicating that nearby points project to nearby points. Randomness disrupts this convergence.

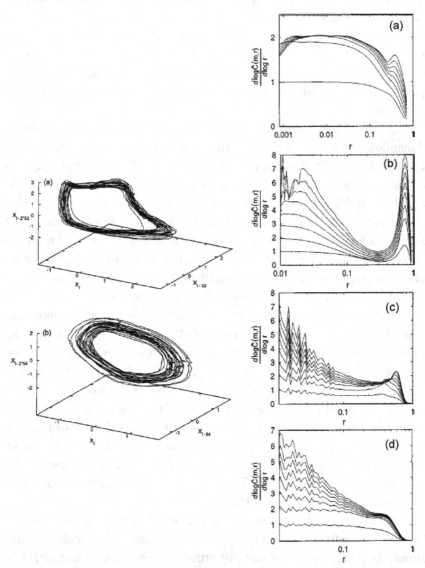

Figure 17.3.1. Left: reconstructed state-space attractors for essential tremor (top) and Parkinsonian tremor (bottom). Right: Slope of correlation integrals for: a) Lorenz system (low-dimensional chaos), b) noisy van der Pol oscillator, c) essential tremor, d) Parkinsonian tremor. (Reused with permission from J. Timmer, *Chaos*, 10, 278 (2000). Copyright 2000, American Institute of Physics.)

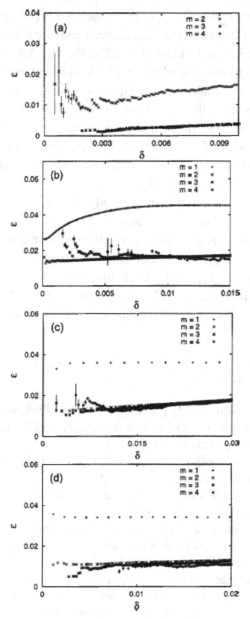

Figure 17.3.2. Results of δ-ε method applied to: a) Lorenz system (low-dimensional chaos), b) noisy van der Pol oscillator, c) essential tremor, d) Parkinsonian tremor. (Reused with permission from J. Timmer, Chaos, 10, 278 (2000). Copyright 2000, American Institute of Physics.)

As seen in Fig. 17.3.2a, for the chaotic Lorenz system (and embedding dimension greater than 2), ε indeed converges to near zero as δ decreases, at the left end of the graph. For the van der Pol oscillator (panel b), essential tremor (panel c), and Parkinsonian tremor (panel d), projected distance ε did *not* converge to near zero as initial distance δ decreased, suggesting the presence of significant randomness. Furthermore, the δ-ε pattern within each system was unchanged for embedding dimensions of 2 or more, suggesting that although noise is present, a second-order (two-dimensional) embedding is sufficient to capture the characteristics of the noise.

These and other tests applied in the study all lead to the conclusion that, contrary to the earlier report, Parkinsonian tremor does not appear to be chaotic. Indeed it is in many ways indistinguishable from essential tremor, which is a disappointing conclusion as far as clinical diagnosis. These tremors instead appear to be nonlinear (as determined in the previous study, and indicated by spectral properties) and random (as determined by lack of convergence of the correlation dimension and the δ-ε algorithm).

References for Chapter 17

H Bergman, G Deuschl (2002) Pathophysiology of Parkinson's disease: from clinical neurology to basic neuroscience and back. *Movement Disorders* 17 Suppl 3:S28-S40.

EN Bruce (2001) Biomedical Signal Processing and Signal Modeling. New York: John Wiley & Sons.

G Deuschl, M Lauk, J Timmer (1995) Tremor classification and tremor time series analysis. *Chaos* 5:48-51.

C Gantert, J Honerkamp, J Timmer (1992) Analyzing the dynamics of hand tremor time series. *Biological Cybernetics* 66:479-484.

AD Pinto, AE Lang, R Chen (2003) The cerebellothalamocortical pathway in essential tremor. *Neurology* 60:1985-1987.

M Shelhamer (1997) On the correlation dimension of optokinetic nystagmus eye movements: computational parameters, filtering, nonstationarity, and surrogate data. *Biological Cybernetics* 76:237-250.

J Timmer, C Gantert, G Deuschl, J Honerkamp (1993) Characteristics of hand tremor time series. *Biological Cybernetics* 70:75-80.

J Timmer, S Haussler, M Lauk, CH Lucking (2000) Pathological tremors: deterministic chaos or nonlinear stochastic oscillators? *Chaos* 10:278-288.

Chapter 18

Case study – Neural dynamics and epilepsy

There is an absolutely enormous and rapidly expanding literature on the application of nonlinear dynamics in neuroscience, and in particular to the analysis and prediction of epileptic seizures. No single chapter (or even entire book) can hope to survey all of this material and do it justice. We focus here on the issue of seizure prediction – is there some feature of an EEG record that can be measured that will indicate, ahead of time, that an epileptic seizure is about to occur? The benefits of such an ability, made available in an implanted device, are obvious for patient health and welfare. While there are many approaches to this problem, we discuss two, one with some identified problems and the other seemingly more promising. This problem is by no means solved.

We begin with a general introduction and a brief survey of some early dynamical approaches to EEG analysis.

18.1 Epilepsy background

Epilepsy is a major worldwide health problem with debilitating consequences for length and quality of life (see for example reviews in Iasemidis 2003, Lopes da Silva *et al.* 2003). It is manifest as recurring seizures, during which there can be changes in cognitive state, undesired automatic movements, and convulsions. It can be congenital, or acquired through disease or trauma.

The accepted cause of a seizure is abnormal synchronous neuronal firing in one or several cortical areas. The resulting *ictal* spikes in the EEG are the result of this hypersynchronized cortical activity. The epileptogenic focus might be localized to a small area or more widely distributed, and may be due to the coupling of spatially disparate neuron pools. This leads to a general decrease in the "degrees of freedom" (or "dynamical dimension") of neural activity.

The transition from normal brain state to seizure can be smooth and continuous, or sudden. A smooth transition means that there is a period of time when, in principle, neural changes are taking place that might be detectable by the proper monitoring procedure. The presence in some cases of an aura – a subjective perception that precedes a seizure – also supports the idea that there are neural changes preceding the seizure itself.

Possible seizure interventions include the administration of appropriate drugs, surgery, or electrical stimulation to disrupt the hypersynchronization. A timely warning of an impending seizure would provide sufficient time for drug administration or electrical stimulation, especially from implanted devices. Long-term monitoring for dynamical changes could also provide a means of assessing chronic medication levels.

Nonlinear dynamics provides a conceptual framework for understanding and recognizing spatial-temporal dynamics and patterns as might result from the type of neural synchronization that leads to a seizure. An appealing conceptual framework posits that there are transitions to and from chaotic behavior as the effective number of degrees of freedom changes. While the new approaches brought about by the application of dynamical thinking to this field may allow for the detection of imminent seizures even without a complete understanding of the neural mechanisms, it is likely that these methods will lead to enhanced understanding as well.

18.2 Initial dynamical studies

Babloyantz and Destexhe (1986) were apparently the first to analyze epileptic EEG using the nonlinear methods that we consider. They showed a reduction in the correlation dimension from 4.05 during sleep to 2.05 during epileptic seizure. The study is interesting for its foresight, despite some shortcomings in this early application of the computational procedures (some noted soon thereafter (Dvorak & Siska 1986)).

Shortly afterward, a study was performed on EEG records from normal human subjects (Rapp *et al.* 1989), which found median correlation dimensions of 3.4-4.3, 4.8±0.9, and 4.8 ±0.2 for subjects at rest, counting upward by two, and counting backward by seven, respectively. These results were almost too good, showing a clear relationship of EEG "complexity" to the prevailing mental task. Indeed, a later study (Theiler & Rapp 1996) reexamined these results, and did find them too good to be true. The later study imposed stricter requirements on the dimension computations (as described in Chapter 5), including the Theiler window to avoid temporal correlations, a limit on the maximum allowed dimension value based on the number of attractor points, and a minimum acceptable scaling region corresponding to a factor of five from minimum to maximum distance values. With these new standards, none of the original EEG records gave a finite dimension value. The scaling region standards were then relaxed, and dimensions were compared to those of phase-randomized surrogates. With this test, the hypothesis of linearly correlated Gaussian noise could be rejected for a subset of the EEG records. However, given the relaxed criteria for these dimension computations, the evidence for low-dimensional deterministic dynamics in these normal EEG samples is equivocal at best.

Similarly, another early study (Frank *et al.* 1990) found a correlation dimension of about 5.6 in a single human epilepsy sample. This result was then examined in more detail with surrogate data (Theiler 1995, this example described briefly in section 6.10 of Chapter 6). Two surrogates were tested: phase-randomized, and random shuffling of data segments each consisting of a spike and the subsequent "wave" before the next spike. In this case the dimension estimates did not converge with increasing embedding dimension, for either the original data or the surrogates, again providing an equivocal result.

A more promising early result was found during induced seizures in rat hippocampus (Pijn *et al.* 1991). Correlation dimensions were high or unobtainable (no convergence) during rest or locomotion, with no difference from phase-randomized surrogates. The induced seizures, however, produced a low dimension (2-4) which was significantly different from that of the corresponding surrogates. The surrogate result in particular is important since it shows that a change in the frequency spectrum itself does not produce the change in dimension with seizure.

Despite the lack of firm results from many of these pioneering early studies, they did lay the groundwork for further studies by showing the great care that must be taken in computing dimensions (and other dynamical measures) from EEG data. This spurred later studies in several ways: claims were made more circumspect, procedures were made more thorough, and less ambitious goals were set in cases where only relative or comparative measures are sufficient. Prediction of imminent epileptic seizures is an obvious case of the last point. Prediction of course leads to the possibility of therapeutic intervention, even possibly the use of chaos control strategies as discussed in Chapter 12. Before there is real success in this endeavor, however, the confounding issues of noise, natural changes in physiological state, and difficulty in identifying a clean reference state, need to be addressed in detail.

18.3 Dimension as a seizure predictor

Based on some of the early studies on changes in EEG dimension in different brain states, it was natural to focus attention on dimension changes that might predict the onset of a seizure. The problem lies in making a dimension estimate rapidly and with limited data, and then tracking temporal changes in the dimension; these are requirements for a practical real-time (rather than *post hoc*) seizure predictor.

One approach to this problem (Lehnertz & Elger 1995, 1998) is to define an "effective dimension," D^{eff}, based on the correlation dimension, but greatly simplified. The process begins by reconstructing the attractor with embedding dimensions $M=1$ and $M=30$. The correlation integrals and their slopes are found in the usual way, as described in Chapter 5. Automated determination of the scaling region (range of distances r over which the slopes of the correlation integrals are constant) is the key to this approach. First, the upper bound r_u is defined at the value of r where the slope of the correlation integral decreases below a value of ~1, for

$M=1$ (this occurs when the criterion distance r has encompassed the entire data set). Then, starting at r_u, the slope of the correlation integral at $M=25$ (where presumably the attractor has been properly reconstructed) is examined at decreasing values of r, until the slope at a particular value of r deviates by more than 5% from the value at r_u. This defines the lower bound r_l, and the slopes at those value of r between r_l and r_u (for $M=25$) are averaged to obtain D^{eff}. If these steps do not yield a value, then D^{eff} is set to an upper bound of 10.

The quantity D^{eff} was computed for consecutive half-overlapping EEG segments, each 30 sec in duration (at 173 Hz sampling rate, $N=5190$ points per segment). Changes in D^{eff} were monitored over time, looking for a reduction of large magnitude and duration as reflecting a reduction in neural "complexity." An example is shown in Fig. 18.3.1, which shows D^{eff} tracked over 35 minutes at three electrode sites. Large and consistent reductions are seen several minutes before seizure onset at 24 minutes (dashed line).

These changes in D^{eff} were distributed across recording sites, but were greatest near the epileptic focus, which shows the distributed nature of the neuronal recruitment leading to a seizure, and suggests that this approach could be used to help localize the site of epileptogenesis.

The authors take pains to make clear that D^{eff} should not be construed as a dimension estimate *per se*, but only as a relative measure related to "dynamical complexity." The computations leading to D^{eff} are somewhat arbitrary, yet objective, and could easily be incorporated into an automated seizure prediction algorithm. The transient decrease in D^{eff} that precedes a seizure indicates a loss of complexity, a reduction in the prevailing "degrees of freedom," consistent with many other studies which generally show a reduction in "complexity" with pathology. In this specific case it is likely due to recruitment and synchronization of separate neuronal populations.

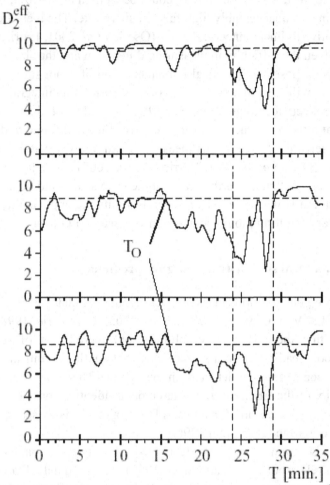

Figure 18.3.1. Effective dimension versus time, from three recording sites. D^{eff} exhibits large and sustained reductions several minutes prior to seizure onset at the first dashed line. (Reprinted with permission from Lehnertz & Elger, *Phys Rev Lett* 80:5019-5022, 1998. Copyright 1998 by the American Physical Society.)

This promising result was later called into question by several investigators. Aschenbrenner-Scheibe *et al.* (2003) applied the same method to longer-term recordings (24-hour), and found a serious degradation in prediction ability caused by now having to deal with large variations due to nonstationarity from natural changes in physiological state (sleep, rest, wake). Sensitivity could be as high as 30-40%, but only by allowing an unacceptably high false-positive rate. The measure D^{eff} is also sensitive to linear autocorrelations (Osorio et al. 2001, Harrison et al. 2005). Given the difficulty in measuring dimension and the shortcuts that need to be imposed for practical implementation, it is not surprising that some flaws will exist in approaches based on correlation integrals.

In a related technique (Lerner 1996), the value of the correlation integral at a fixed distance r was tracked over time and found to decrease before a seizure. However, this measure is extremely sensitive to signal amplitude (Osorio et al. 2001, Harrison et al. 2005), since larger signals will have points that are farther apart in the state space, and no amplitude normalization is performed. This severely limits the applicability of this method for long-term monitoring of nonstationary EEG data.

18.4 Dynamical similarity as a seizure predictor

A more general approach is to track dynamical changes leading to a seizure (Le Van Quyen et al. 2001a, 2005, Martinerie 1998). This assumes that there is a slow sub-threshold recruitment of separated neural pools leading up to a seizure, and that subtle dynamical changes as this occurs can be detected in the EEG. This does not assume stationarity of the EEG, but rather attempts to identify nonstationarity as reflecting a change in dynamics. (The approach is based on ideas developed in Manuca & Savit 1996.)

The algorithm was first developed on epileptic patients with implanted electrodes (Le Van Quyen 1999). First, a quiet EEG epoch is selected as a reference, distant in time from any seizure activity. A time-delay reconstruction is used to create a reference attractor, but instead of using the raw EEG samples, the data consist of time intervals between threshold crossings of the EEG signal (positive-going zero-crossings). This produces a data set that is roughly analogous to a Poincaré section, is insensitive to signal amplitude and noise, and provides a dramatic reduction in the amount of data. (The process of attractor reconstruction from such time-interval data was discussed in Chapter 4 (see also Sauer

1994).) An embedding dimension of $M=16$ was used, using the time intervals as data values.

At this point a singular value decomposition was applied, in order to reduce the dimensionality of the reconstructed attractor, by identifying and retaining only those directions in the state space along which significant activity occurs. (See brief discussion in section 5.9 of Chapter 5.) Understanding of this step is not critical to understanding the overall procedure; it is largely introduced with a view to the reduction of computational burden, for eventual real-time monitoring applications.

Having identified a reference attractor composed of the points $X(S_{ref})$, a set of test segments is found, by selecting consecutive non-overlapping EEG records, each 25 sec in duration. For each test segment, the attractor is reconstructed as above, with time-interval data, forming the points $X(S_t)$. (These attractors are processed with the same singular value decomposition as for the reference segment.)

At this point, the *cross-correlation integral* is found, for each test segment compared to the reference segment. This is a modification of the standard correlation integral (Chapter 5) used in the computation of the correlation dimension. It is defined as:

$$C(S_{ref}, S_t, \varepsilon) = \frac{1}{N_{ref}^2 N_t^2} \sum_{i=1}^{N_{ref}} \sum_{j=1}^{N_t} \mathbf{U}(\varepsilon, |X_i(S_{ref}) - X_j(S_t)|) \quad (i \neq j)$$

$$\mathbf{U}(\varepsilon, |X_i(S_{ref}) - X_j(S_t)|) = \begin{cases} 1 & \text{if } |X_i(S_{ref}) - X_j(S_t)| < \varepsilon \\ 0 & \text{otherwise} \end{cases}$$

The value of the integral, $C(\cdot)$, provides a measure of the "dynamical similarity" between the two attractors $X(S_{ref})$ and $X(S_t)$. It is the likelihood of finding attractor points from the reference and test sets that are closer to each other than the reference distance ε (ε is set to 30% of the cumulative distribution of the inter-point distances in the reference set). (N_{ref} and N_t are the number of points in a given reference and test attractor, respectively.)

In practice, the value of the integral is normalized to give a similarity index between 0 and 1, with 1 indicating high similarity:

$$\gamma(S_{ref}, S_t) = \frac{C(S_{ref}, S_t)}{\sqrt{C(S_{ref}, S_{ref})C(S_t, S_t)}}.$$

For the validation study (Le Van Quyen 1999), the algorithm was tested on 13 patients with middle temporal lob epilepsy. After selecting the reference segment, the first ten 25-second EEG segments were analyzed to provide a baseline for non-seizure variations from the reference: the mean and standard deviation of the resulting ten values of γ were used to evaluate the statistical significance of γ excursions (reductions in similarity relative to the reference) below this level in subsequent EEG test segments.

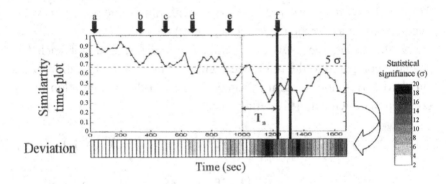

Figure 18.4.1. Results of analysis of dynamical similarity applied to a single epileptic EEG recording. Graph shows time course of similarity measure γ, with large decreases (loss of similarity) indicated by arrows. Seizure begins at arrow f. Based on statistical criteria in cited study, seizure is anticipated by interval T_a. (Figure 1, bottom, from Le Van Quyen et al. 1999, with permission.)

A typical result is shown in Figure 18.4.1, which graphs the value of γ for a single subject for several minutes before the onset of a seizure (at arrow f). Each of the arrows labeled a-e indicates a sharp reduction in similarity index γ, until there is a large and prolonged reduction immediately preceding the seizure. Using statistical criteria from the validation study, these excursions would have indicated the impending seizure more than 200 seconds in advance, which is sufficient time for an appropriate intervention. In general, seizure predictions were made several minutes in advance. Furthermore, changes in γ were present at several different recording sites, indicating that neural assemblies distributed over several cortical regions were being jointly recruited in the period leading up to the seizure. This opens the possibility of using such spatial information explicitly in extensions of this procedure.

These promising results were later verified with surface EEG recordings (Le Van Quyen 2001). This algorithm, like any of its sort, needs to be validated by appropriate surrogates and by ascertaining how often the critical reductions in similarity index occur very far in time from a seizure (false positives), especially since EEG records used in the cited study were always within 20 minutes of a seizure. Further investigation should also determine what specific dynamical aspects are measured by the similarity index (the basis on zero-crossing times suggests that it might be sensitive to frequency changes).

18.5 Validation with surrogates, comparison of procedures

One approach to statistical validation of a given prediction algorithm (Andrzejak *et al.* 2003) is to generate *seizure-time surrogates*, by discarding the actual seizure times in a long EEG record and "pretending" that the seizures actually occurred at other random times. Any viable prediction scheme should *not* predict these surrogate seizures since they are *not* (in general) preceded by a true pre-seizure state. The null hypothesis for this surrogate test is: The transition from the interictal to the ictal state is an abrupt phenomenon. An intermediate preictal state does not exist. ("Ictal" refers to a seizure; "interictal" is the time period between seizures.)

A more sophisticated extension of this idea is a *measure profile surrogate* (Kreuz *et al.* 2004). The times of the seizures are maintained. The *measure profile* is randomized. The measure profile is the time series of values that is produced by the prediction algorithm (such as effective dimension or dynamical similarity, in the examples given previously). The randomization is carried out so as to maintain certain desirable statistical properties of the measure profile, such as temporal correlations. Whatever statistical criterion has been developed to determine when a significant change in the measure has taken place (e.g., reduction in dimension by a certain amount for a certain duration), is applied to the rearranged measure profile. The detection level should be significantly worse for the surrogate than for the original measure profile if the measure is indeed a good one for predicting seizure onset.

Statistical considerations for a viable seizure prediction algorithm, such as allowable false-positive rate, desirable prediction horizon, and related parameters, have been incorporated into a quantitative "seizure prediction characteristic" for testing of proposed methods (Winterhalder *et al.* 2003, Maiwald *et al.* 2004).

Several investigators have conducted studies that compare the performance of different seizure prediction methods (e.g., Jerger *et al.* 2001, Mormann *et al.* 2005). A general finding is that most of the procedures can be made to work to some extent, although such practical considerations as false-positive rate and false-negative rate are still problematic. The methods also perform better the more specifically tailored they are to a given situation – by allowing a baseline reference measure for each data channel and each seizure record, for example.

Another general finding is that linear measures perform the same or better than nonlinear measures. This is surprising, given the unquestionably nonlinear nature of neural dynamics. The disappointing performance of the nonlinear methods might be because the dynamical nonlinearities in the data stream are too complex to be picked up by these methods, so that the nonlinear methods are really measuring linear properties. It may be that much better and more extensive data will be needed in order for the nonlinear methods to show their full potential (Jerger *et al.* 2001). It is also possible that the great complexity of normal neural functioning leads to "averaging out" of largely uncorrelated nonlinear aspects, leaving more well-behaved linear properties (see end of section 20.2 for a similar effect in epidemiological modeling). This possibility would suggest that the reductions in "complexity" identified by some seizure-prediction methods might reflect a transition from high-dimensional linearly correlated noise to low-dimensional nonlinear determinism, as neuronal assemblies become synchronized. It is also worth noting that among the most promising of the methodologies are those that detect phase-synchronization of EEG patterns in different cortical areas (e.g., Le Van Quyen 2005), which is not necessarily an inherently nonlinear phenomenon.

These studies do, however, show without doubt that there is a detectable pre-seizure state, which is not an insignificant finding. Without such a state the entire endeavor of predicting seizure onset would be impossible.

References for Chapter 18

RG Andrzejak, F Mormann, T Kreuz, C Rieke, A Kraskov, CE Elger, K Lehnertz (2003) Testing the null hypothesis of the nonexistence of a preseizure state. *Physical Review E* 67:010901-1:4.

R Aschenbrenner-Scheibe, T Maiwald, M Winterhalder, HU Voss, J Timmer, A Schulze-Bonhage (2003) How well can epileptic seizures be predicted? An evaluation of a nonlinear method. *Brain* 126:2616-2626.

A Babloyantz, A Destexhe (1986) Low-dimensional chaos in an instance of epilepsy. *Proceedings of the National Academy of Sciences of the USA* 83:3513-3517.

I Dvorak, J Siska (1986) On some problems encountered in the estimation of the correlation dimension of the EEG. *Physics Letters A* 118:63-66.

GW Frank, T Lookman, MAH Nerenberg, C Essex, J Lemieux, W Blume (1990) Chaotic time series analyses of epileptic seizures. *Physica D* 46:427-438.

MA Harrison, I Osorio, MG Frei, S Asuri, YC Lai (2005) Correlation dimension and integral do not predict epileptic seizures. *Chaos* 15:33106-1:15.

D Iasemidis (2003) Epileptic seizure prediction and control. *IEEE Transactions on Biomedical Engineering* 50:549-558.

KK Jerger, TI Netoff, JT Francis, T Sauer, L Pecora, SL Weinstein, SJ Schiff (2001) Early seizure detection. *Journal of Clinical Neurophysiology* 18:259-268.

T Kreuz, RG Andrzejak, F Mormann, A Kraskov, H Stögbauer, CE Elger, K Lehnertz, P Grassberger (2004) Measure profile surrogates: a method to validate the performance of epileptic seizure prediction algorithms. *Physical Review E* 69:061905-1:9.

K Lehnertz, CE Elger (1995) Spatio-temporal dynamics of the primary epileptogenic area in temporal lobe epilepsy characterized by neuronal complexity loss. *Electroencephalography and Clinical Neurophysiology* 95:108-117.

K Lehnertz, CE Elger (1998) Can epileptic seizures be predicted? Evidence from nonlinear time series analysis of brain electrical activity. *Physical Review Letters* 80:5019-5022.

K Lehnertz, F Mormann, T Kreuz, RG Andrzejak, C Rieke, P David, CE Elger (2003) Seizure prediction by nonlinear EEG analysis. *IEEE Engineering in Medicine and Biology Magazine.* 22:57-63.

DE Lerner (1996) Monitoring changing dynamics with correlation integrals: case study of an epileptic seizure. *Physica D* 97:563-576.

M Le Van Quyen, J Martinerie, M Baulac, F Varela (1999) Anticipating epileptic seizures in real time by a non-linear analysis of similarity between EEG recordings. *Neuroreport* 10:2149-2155.

M Le Van Quyen, J Martinerie, V Navarro, M Baulac, FJ Varela (2001a) Characterizing neurodynamic changes before seizures. *Journal of Clinical Neurophysiology* 18:191-208.

M Le Van Quyen, J Martinerie, V Navarro, P Boon, M D'Have, C Adam, B Renault, F Varela, M Baulac (2001b) Anticipation of epileptic seizures from standard EEG recordings. *Lancet* 357:183-188.

M Le Van Quyen (2005) Anticipating epileptic seizures: from mathematics to clinical applications. *Comptes Rendus Biologies* 328:187-198.

FH Lopes da Silva, W Blanes, SN Kalitzin, J Parra, P Suffczynski, DN Velis (2003) Dynamical diseases of brain systems: different routes to epileptic seizures. *IEEE Transactions on Biomedical Engineering* 50:540-548.

T Maiwald, M Winterhalder, R Aschenbrenner-Scheibe, HU Voss, A Schulze-Bonhage, J Timmer (2004) Comparison of three nonlinear seizure prediction methods by means of the seizure prediction characteristic. *Physica D* 194:357–368.

R Manuca, R Savit (1996) Stationarity and nonstationarity in time series analysis. *Physica D* 99:134-161.

J Martinerie, C Adam, M Le Van Quyen, M Baulac, S Clemenceau, B Renault, FJ Varela (1998) Epileptic seizures can be anticipated by non-linear analysis. *Nature Medicine* 4:1173-1176.

F Mormann, T Kreuz, C Rieke, RG Andrzejak, A Kraskov, P David, CE Elger, K Lehnertz (2005) On the predictability of epileptic seizures. *Clinical Neurophysiology* 116:569–587.

I Osorio, MA Harrison, YC Lai, MG Frei (2001) Observations on the application of the correlation dimension and correlation integral to the prediction of seizures. *Journal of Clinical Neurophysiology* 18:269-274.

JP Pijn, J Van Neerven, A Noest, FH Lopes da Silva (1991) Chaos or noise in EEG signals; dependence on state and brain site. *Electroencephalography and Clinical Neurophysiology* 79:371-381.

PE Rapp, TR Bashore, JM Martinerie, AM Albano, ID Zimmerman, AI Mees (1989) Dynamics of brain electrical activity. *Brain Topography* 2:99-118.

T Sauer (1994) Reconstruction of dynamical systems from interspike intervals. *Physical Review Letters* 72:3811-3814.

J Theiler (1995) On the evidence for low-dimensional chaos in an epileptic electroencephalogram. *Physics Letters A* 196:335-341.

J Theiler, PE Rapp (1996) Re-examination of the evidence for low-dimensional, nonlinear structure in the human electroencephalogram. *Electroencephalography and Clinical Neurophysiology* 98:213-222.

M Winterhalder, T Maiwald, HU Voss, R Aschenbrenner-Scheibe, J Timmer, A Schulze-Bonhage (2003) The seizure prediction characteristic: a general framework to assess and compare seizure prediction methods. *Epilepsy & Behavior* 4:318-325.

Chapter 19

Case study – Cardiac dynamics and fibrillation

Given the prevalence of cardiovascular disease and its consequences, the need for improved diagnostic tools is obvious. The ability to predict the onset of imminent fibrillation, for example, would have great public health benefits. After presentation of some preliminary material on heart-rate variability and its dynamics, we discuss two nonlinear dynamics approaches to the prediction of imminent fibrillation. A definitive presentation of this work is difficult, since much of it seems to have gone "underground" and become proprietary.

19.1 Heart-rate variability

Analysis of heart-rate variability (HRV) is a simple, non-invasive method for monitoring changes in cardiac function as a result of exercise, aging, pathology, or stress (Tsuji *et al.* 1996, Huikuri & Mäkikallio 2001). Although the resting heart rate is approximately constant, there is some beat-to-beat variation, and larger variations due to the processes just listed. These variations come about largely though the extensive autonomic innervation of the heart (Pumprla *et al.* 2002). Parasympathetic activity expressed through acetylcholine causes a rapid and relatively short-acting decrease in rate, while sympathetic activation through noradrenaline increases the rate and acts more slowly. The difference in response times of the two subsystems enables some differentiation of their influence, through frequency analysis of HRV. There are also modulations in rate due to respiration and blood pressure, as well as changes in posture and other activities. These many and varied influences lead to a rich dynamical complexity in beat-to-beat control.

Degradation of these normal modulatory influences can have adverse health consequences, and can in some cases be detected in altered measures of HRV: "Lower variability is often an indicator of abnormal and insufficient adaptability of the autonomic nervous system" (Pumprla et al. 2002). "Abnormalities of autonomic input to the heart, which are indicated by decreased indices of HRV, are associated with increased susceptibility to ventricular arrhythmias" (Stein et al. 1994).

In particular, the promise of HRV to predict upcoming life-threatening ventricular fibrillation (rapid irregular and ineffectual heart activity) has great appeal and drives much current research. "The relationship between reduced HRV and prognosis in patients with cardiovascular disease may in part be related to the role of the autonomic nervous system in regulating myocardial electrical stability. Sympathetic activation favours the onset of life-threatening ventricular tachyarrhythmias, whereas parasympathetic activation exerts a protective and antifibrillatory effect, and abnormalities of HRV reflecting adverse changes in autonomic activity have been demonstrated immediately prior to the onset of ventricular tachyarrhythmias" (Pumprla et al. 2002).

HRV measures are based on inter-beat intervals as derived from EKG (electrocardiography) recordings of heart activity. The large and predominant QRS complexes of the EKG occur with each beat, and it is usually a simple matter to detect the R wave and so derive R-R intervals (inter-beat intervals).

Conventional measures of HRV include standard deviation, frequency content in various ranges, and other more sophisticated measures derived from linear principles and random-process theory, and which generally assume stationarity. Many HRV studies until recently have been retrospective, and show that these conventional measures are useful as long-term predictors of mortality in patients who have already had a heart attack, while their utility for warning of imminent fibrillation is less certain (Vybiral et al. 1993). Given the complex nature of the subsystems that determine heart rate, and their interactions, it is not at all unlikely that HRV will contain a significant amount of nonlinear dynamical activity. Nonlinear measures of HRV, therefore, might be expected to have great promise as indicators and predictors of cardiac health and disease.

19.2 Noisy clock or chaos?

An early study of the dynamics of normal cardiac inter-beat intervals (Babloyantz & Destexhe 1988) found correlation dimensions from 3.6 to 5.2, using the entire EKG signal (not inter-beat intervals). This study also used attractor reconstruction and Poincaré sections to make a case for deterministic and possibly chaotic structure. The basic question addressed in the study is a simple one, but one which has continued to occupy many subsequent researchers: is the heart rhythm a noisy clock, or is it chaotic? In the first case, the dynamics are classically deterministic, and variability is due to imposed noise. In the second case, the dynamics are deterministic but the variability arises from the fact that the dynamics are chaotic; the variability is inherent in the dynamics. This key question underlies many of the physiological applications of the methods presented in this book.

Resolution of the question is complicated by the fact that there is undoubtedly some variation introduced by measurement procedures, as well as by physiological factors not inherent to the heart itself. An example of the latter is the *respiratory sinus arrhythmia*, which is the normal variation in heart rate with respiration, mediated through sympathetic activity in the vagus nerve (this small increase in rate upon inspiration can often be detected by monitoring the pulse on the wrist while inhaling deeply and then holding the breath).

A slightly more modern study along the same lines came just a few years later (Kaplan & Cohen 1990). These authors were specifically interested in the question of whether or not fibrillation is chaotic (in the technical sense: irregular but deterministic, with finite non-integer dimension). This is a reasonable question, based on the simple fact that ventricular fibrillation "looks" turbulent and "chaotic." Dimensions from phase-randomized surrogates and from fibrillation were indistinguishable: there was no convergence up to an embedding dimension of 15, suggesting a large role for randomness in this situation.

A subsequent study (Govindan *et al.* 1998) also examined the entire EKG signal, using dimension analysis, with surrogates. These results suggest that EKG may have some deterministic dynamics and might even be chaotic, in normals as well as in various pathologies. There was not, however, much discriminatory power between the patient groups, except that the dimension for ventricular fibrillation was much higher than for the other groups. Forecasting results were consistent with these findings: forecasting quality decayed rapidly with time step. Quality was good for times less than that of an inter-beat interval, which questions the relevance of this study for understanding the more extensive inter-beat interval analyses, such as those presented in the next section.

Heart transplantation entails a major disruption of descending neural influences on heart rate, and so measurement of HRV in transplant patients is one way to assess the role of these pathways. Bogaert *et al.* (2001) measured the correlation dimension of inter-beat intervals from normal subjects and transplant patients, finding little difference between the two groups, although the dimension values did not converge for most transplant patients (dimensions 2.12-5.53 in healthy normals, 2.10-5.6 in patients). Surrogates had higher dimensions than the actual data in both cases, showing that there is some nonlinear structure. The data records in this study were relatively short (700-850 intervals), which although not explicitly violating the 2 log(N) rule for dimensions, do come close. It is not clear how long after transplantation these patients were tested.

An different study (Guzzetti *et al.* 1996) measured HRV in patients six months after transplantation. This study also showed a reduction in dimension in patients, and a simpler structure in the first-return maps. Dimension values were still quite large (from about 7 in normals to about 4 in transplant patients). These data sets were large (about 20,000 points), but this apparently introduced nonstationarity in the data records which may have contributed to the high dimension values.

In contrast, using a different algorithm that can determine instantaneous dimensions and track temporal changes (Meyer *et al.* 1996, see section 19.4 below), dimensions in transplant patients were found to be much lower. This method is less influenced by nonstationarity, which invariably increases the apparent dynamics of long EKG records. Average dimension values in this study were reduced after transplantation, from 5.4 in normal control subjects to 1.1±0.1 in patients within a year of transplantation, with gradual increases thereafter. In one patient the mean value was 3.7 within 2 weeks of transplant. These results indicate that, indeed, the time at which the data are recorded matters a great deal, as immediate loss of innervation leads to a high dimension (possibly due to increased randomness or reorganization as noted below), after which recovery of normal innervation patterns is reflected in a gradual (but incomplete) return of the dimension to lower levels.

Some suggestions for reliable automated computation of the correlation dimension from entire EKG signals have been made (Casaleggio & Bortolan 1999).

19.3 Forecasting and chaos

An extensive study of HRV dynamics (Lefebvre *et al.* 1993) approached the issue from the forecasting perspective: can nonlinear forecasting be used to distinguish between healthy and pathologic populations? Four groups of subjects were examined: young healthy (YH), older healthy (OH), patients who had had a heart attack and were being treated with beta-blocker medication (B), and heart-attack patients not being treated with this medication (NB).

Data sets were drawn from 45 minutes of resting EKG, taking 1001 beats from what appeared to be a steady-state rhythm. First differences were taken to remove any gradual trends. An attempt was made to measure the correlation dimension, but there was no clear convergence of the dimension estimates with increasing embedding dimension; there was some suggestion of a dimension between 5 and 6 but it was not robust. Dimension values this high are also questionable with this few data values (see Chapter 5).

One interesting aspect of the dimension computations in this study is worthy of comment. The data were sampled with a time resolution of 1 msec. This means that inter-beat interval values are subject to noise (round-off error) due to this finite resolution. In the time-delay reconstruction, therefore, each component k of each M-dimensional attractor point has an error (standard deviation, sd) on the order of 1 msec, which we will term ε. Recall that the Euclidean distance between two attractor points $y(i)$ and $y(j)$ is:

$$dist = \sqrt{\sum_{k=1}^{M}[y^k(i) - y^k(j)]}.$$

Assume that we have two points on the attractor that are so close together that this noise term dominates the distance computation. Since these components are subtracted term-by-term in the distance computation, the sd of each term is just the sd of the sum of two random variables, each with an sd of ε: this value is $2^{1/2}\varepsilon$ (the variances add directly, and a square root is taken to obtain sd). There are M components in computing the distance between these two points, and so the uncertainty (sd) of the distance is:

$$dist\ s.d. = \sqrt{\sum_{k=1}^{M}[\sqrt{2}\varepsilon]^2} = \sqrt{M[\sqrt{2}\varepsilon]^2} = \sqrt{2M\varepsilon}.$$

These values were plotted on the correlation integral curves, at each value of M, in order to verify that the scaling region was not affected by this source of noise. This is an excellent example of the kind of numerical control that should accompany any such investigation.

The forecasting method of Sugihara and May (see Chapter 7) was then applied to the interbeat-interval data, with very slight modifications. Results are summarized in Fig. 19.3.1. As usual, there is a decrease in forecasting ability with time step, for all subject groups. The first-step forecasting ability and forecasting decay rate (slope of correlation coefficient versus time step) were then measured.

First-step forecasting for the YH group (panel a) was better than that for the OH group (panel b) (mean correlation coefficient 0.63 vs. 0.47). This is surprising, given the increasing evidence for a reduction in dynamical complexity (and hence an increase in forecasting ability) with age.

On the other hand, there was a difference in forecasting decay rate between the OH group and both patient groups, and this difference was statistically significant in the case of comparison with the non-beta-blocker group (panel d). The decay in forecasting quality was more rapid for the OH group, which suggests that the patients have, on average, more low-dimensional deterministic structure. This is a central finding of the study.

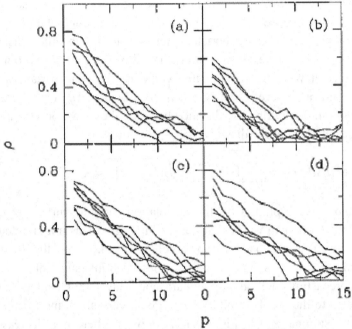

Figure 19.3.1. Nonlinear forecasting results: correlation coefficient between actual and forecast values (ρ) versus forecasting time step (p), for four subject groups: a) young healthy, b) older healthy, c) heart-attack patients taking beta-blockers, d) heart-attack patients not on beta-blockers. (Reused with permission from Lefebvre et al., Chaos, 3, 272 (1993). Copyright 1993, American Institute of Physics.)

One might ask if the differences in forecasting quality between groups were simply due to differences in the standard deviation of the inter-beat intervals, a more conventional measure unrelated to any considerations of nonlinear dynamics. This was apparently not the case, since first-step forecasting quality and sd were not significantly related, nor were decay rate and sd. Thus nonlinear forecasting may be able to detect physiological changes resulting from heart attack which cannot be detected with more standard measures.

Forecasts were then carried out on the data by fitting a linear model to all of the data (an autoregressive model). Surprisingly, this linear fit performed slightly better than the nonlinear forecasting method. This would seem to suggest that there is not significant nonlinearity in the data, and therefore no chaotic dynamics. The good performance of the linear predictor could be due to the fact that it uses all of the data, while nonlinear forecasting performs only small local approximations using small segments of the data in each forecast. It might also be that a more sophisticated nonlinear forecasting algorithm would perform better.

As another attempt to assess nonlinearity in these data, phase-randomized surrogates were then generated. Forecasting was carried out on these data sets. Then, rather than generate an AAFT surrogate (see Chapter 6), the original time series was transformed to a Gaussian signal, by generating a random Gaussian signal and re-ordering the values to match the rank order of the original data. This was done because the surrogates are Gaussian as well. This procedure simulates the hypothesis that the original data were generated by a static monotonic nonlinearity applied to Gaussian data. The surrogates exhibit slightly but consistently less forecasting quality (more randomness) than the transformed original data, again suggesting some degree of nonlinearity in the original data.

The general conclusion of this study is that the inter-beat interval data for these subjects is composed of a small amount of nonlinear determinism, plus noise. This is also indicated by the fact that even the first-step forecast values are rather low (mean correlation coefficient 0.47 to 0.63).

19.4 Detection of imminent fibrillation: point correlation dimension

Our final example is the *point D2* or *PD2* algorithm, mentioned briefly in section 5.7. This computational procedure is based on a dimension algorithm for nonstationary data (Mayer-Kress *et al.* 1988), which was subsequently modified by Skinner *et al.* (1991, 1993, 1994) to track temporal changes in dimension.

Recall that in the conventional correlation dimension computation, many (possibly all) attractor points are used as reference points, and the correlation integrals from all reference points are added together to find a single correlation integral for the entire attractor. The slope of this

more low-dimensional deterministic structure. This is a central finding of the study.

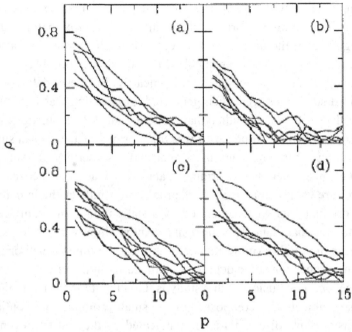

Figure 19.3.1. Nonlinear forecasting results: correlation coefficient between actual and forecast values (ρ) versus forecasting time step (p), for four subject groups: a) young healthy, b) older healthy, c) heart-attack patients taking beta-blockers, d) heart-attack patients not on beta-blockers. (Reused with permission from Lefebvre et al., Chaos, 3, 272 (1993). Copyright 1993, American Institute of Physics.)

One might ask if the differences in forecasting quality between groups were simply due to differences in the standard deviation of the inter-beat intervals, a more conventional measure unrelated to any considerations of nonlinear dynamics. This was apparently not the case, since first-step forecasting quality and sd were not significantly related, nor were decay rate and sd. Thus nonlinear forecasting may be able to detect physiological changes resulting from heart attack which cannot be detected with more standard measures.

Forecasts were then carried out on the data by fitting a linear model to all of the data (an autoregressive model). Surprisingly, this linear fit reaches a value of 0.15x3=0.45. If a scaling region is found under these criteria, then the slope of the correlation integral $C(r)$ over this scaling region is the PD2 value for that reference point.

To assess convergence with M, first define a minimum acceptable M value as 2xD2+1, where D2 is the correlation dimension of the entire attractor. This minimum M is based on the fact that the Takens embedding theorem applies strictly only for embedding dimensions of more than twice the dimension of the attractor (see Chapter 4). The estimated PD2 values at the next four increasing M values must have a standard deviation of less than 0.4 in order to be acceptable, and the final PD2 value is the mean of these values.

This algorithm has been shown to track the instantaneous dimension of a data set formed by concatenating segments with different dynamics. The example in Fig. 19.4.1 shows the concatenation of a sine wave, data from the Lorenz system, more of the sine wave, data from the Hénon system, and more of the sine wave. The dimension of each signal on its own is indicated above each corresponding data segment in the top part of the figure (1.00, 2.06, 1.00, 1.26, 1.00). In the lower panel, the PD2 values are plotted for reference points selected from within each of these data segments, and the mean of the values from within each segment is also shown (1.03, 2.14, 1.03, 1.22, 1.03). The algorithm tracks the changes in dimension with fair accuracy, but the important point is that it adequately tracks dimensional changes over time. Thus even if a specific dimension value is not highly accurate due to poor statistics in the correlation integral, the overall trend of any dimensional changes may be highly useful.

This algorithm has been applied in a number of studies of brain function (Molnar & Skinner 1992, Molnar *et al.* 1997, Aftanas *et al.* 1998, Skinner & Molnar 1999, Sammer 1999, Aftanas & Golocheikine 2002, Toth *et al.* 2004). Here we discuss some of its applications to heart-rate variability, and in particular to the detection of imminent fibrillation.

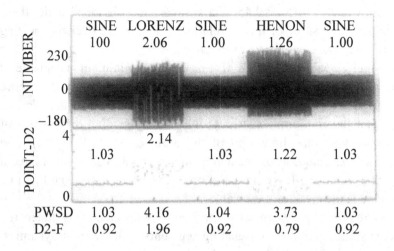

Figure 19.4.1. Demonstration of the PD2 algorithm in tracking the dimension of several concatenated signals. The signals and their individual dimensions are shown at the top, and the PD2 values and their means within a given signal segment are at the bottom.
(From Skinner *et al.* 1993.)

To test the algorithm on physiological data, a study was carried out with controlled occlusion of the coronary artery in pigs (Skinner *et al.* 1991). Epochs of 500 sec were recorded during quiet control conditions, and during occlusions. There were 1200 to 1500 inter-beat intervals in each epoch, which formed the data sets for each sequence of PD2 computations. PD2 values were normally 2.50±0.81, and with complete artery occlusion decreased to 1.58±0.64 in the first minute of ischemia, and then to 1.07±0.18 in the minute preceding ventricular fibrillation. These large and consistent decreases in PD2 only occurred in the time period immediately preceding ventricular fibrillation. Artifacts and transient arrhythmias were automatically rejected, through the scaling criteria described above. Partial artery occlusions produced lesser reductions in PD2 and no fibrillation. Stress and sleep stage also reduced PD2 somewhat, but these factors might also have an effect on vulnerability to fibrillation. PD2 was not significantly correlated with the standard deviation of heart rate.

A subsequent study (Skinner *et al.* 1993) applied the PD2 algorithm to long-term EKG recordings from human subjects. Data were extracted from 24-hour ambulatory recordings; three 6.5-minute segments were selected, one from the start of the recording, one preceding fibrillation, and one in between, and these segments were concatenated for analysis. Patients who experienced (sometimes fatal) ventricular fibrillation during the study were compared to cardiac-high-risk control subjects. As with the animal study above, there was a significant reduction in PD2 immediately preceding fibrillation. Specifically, there were increasingly frequent excursions of PD2 below the value of 1.2 leading up to fibrillation, as shown in Fig. 19.4.2. These episodes were rare in the high-risk patients who did not exhibit fibrillation during the course of the experimental recordings.

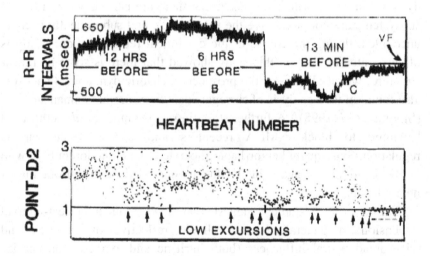

Figure 19.4.2. PD2 values leading up to ventricular fibrillation in one patient. Top: Heart-beat intervals during three segments of a long-term EKG record. The final segment begins 13 minutes before ventricular fibrillation (VF). Bottom: PD2 values over the same time periods as the inter-beat intervals. Low-dimensional excursions become more frequent immediately before fibrillation. (From Skinner *et al.* 1993.)

Subsequent studies were performed in an effort to determine what cardiac characteristics PD2 actually measures. After heart transplant (Meyer *et al.* 1996), PD2 values are near 1.0 in most patients (decreased from a normal value of about 5.4). This low dimension is presumably due to a loss of neurocardiac influence. There is eventually partial recovery of the dimension, suggesting intrinsic reorganization of neural pathways or reinnervation of higher-level autonomic control.

A study of isolated rabbit heart (Skinner *et al.* 1996) found a transient increase in PD2 during ischemia and anoxia, as the cardiac system changed from a strategy of output preservation to one of energy preservation. The temporary dimension increase was thought to reflect increased dynamic complexity during the course of this reorganization of the intrinsic innervation. Analysis of subintervals (intervals between EKG components other than the dominant R waves) showed interesting dynamics as well, which suggested that there are periods when different subsystems are coupled (have the same PD2), and subsequently became uncoupled, during the course of the dynamical reorganization. Analysis of surrogate data sets in this study showed that there is some nonlinear structure, which allows for the prospect of chaotic dynamics. (This was also found in PD2 analysis of surrogate data from normal human subjects (Storella *et al.* 1998).) A further study in rats (Skinner *et al.* 2000) used ketamine to block NMDA receptors and decrease neurocardiac regulation. Subsequent hemorrhagic shock led to a reduction in PD2 with erratic oscillations, again likely due to degraded control from higher neural levels.

These various studies suggest that PD2 (and inter-beat-interval dimension in general) is most likely reflective of complex and widespread neural influences (both intrinsic and extrinsic) on cardiac function. Pathologic reduction in the richness of these descending influences could lead to a reduction in the ability of the organism to respond in an appropriate and healthy manner to external stressors, and this reduction in dynamical complexity would also be manifest in decreased dimension values.

The PD2 algorithm produces a dimension estimate associated with a point in time. For real-time applications, the data that enter into the computation of a PD2 value at a particular point in time must have occurred earlier in time. This limitation does not exist for off-line analysis. It is likely that PD2 estimates would more accurately reflect the underlying attractor if points both preceding and anteceding the reference point could be included in the correlation integral. This can be done by delaying the PD2 estimate until sufficient data are acquired, but this would then cut into the warning time before fibrillation.

It is not clear from these published studies how far in advance of fibrillation the low-dimensional PD2 excursions occur. These studies examined limited data before fibrillation. It could be that significant low-dimensional episodes occur transiently at intervals of days or weeks preceding fibrillation, and these could lead to false-positive indications.

The PD2 algorithm, incidentally, has been patented, which indicates its potential or at least anticipated use in a commercial fibrillation-detection device (US patents 5,720,294 and 5,709,214).

References for Chapter 19

LI Aftanas, SA Golocheikine (2002) Non-linear dynamic complexity of the human EEG during meditation. *Neuroscience Letters* 330:143-146.

LI Aftanas, NV Lotova, VI Koshkarov, VP Makhnev, YN Mordvintsev, SA Popov (1998) Non-linear dynamic complexity of the human EEG during evoked emotions. *International Journal of Psychophysiology* 28:63-76.

A Babloyantz, A Destexhe (1988) Is the normal heart a periodic oscillator? *Biological Cybernetics* 58:203-211.

A Casaleggio, G Bortolan (1999) Automatic estimation of the correlation dimension for the analysis of electrocardiograms. *Biological Cybernetics* 81:279-290.

RB Govindan, K Narayanan, MS Gopinathan (1998) On the evidence of deterministic chaos in ECG: surrogate and predictability analysis. *Chaos* 8:495-502.

S Guzzetti, MG Signorini, C Cogliati, S Mezzetti, A Porta, S Cerutti, A Malliani (1996) Non-linear dynamics and chaotic indices in heart rate variability of normal subjects and heart-transplanted patients. *Cardiovascular Research* 31:441-446.

HV Huikuri, TH Mäkikallio (2001) Heart rate variability in ischemic heart disease. *Autonomic Neuroscience* 90:95-101.

DT Kaplan, RJ Cohen (1990) Searching for chaos in fibrillation. *Annals of the New York Academy of Sciences* 591:367-74.

JH Lefebvre, DA Goodings, MV Kamath, EL Fallen (1993) Predictability of normal heart rhythms and deterministic chaos. *Chaos* 3:267-276.

G Mayer-Kress, FE Yates, L Benton, M Keidel, W Tirsch, SJ Poppl, K Geist (1988) Dimensional analysis of nonlinear oscillations in brain, heart, and muscle. *Mathematical Biosciences* 90:155-182.

M Meyer, C Marconi, G Ferretti, R Fiocchi, P Cerretelli, JE Skinner (1996) Heart rate variability in the human transplanted heart: nonlinear dynamics and QT vs RR-QT alterations during exercise suggest a return of neurocardiac regulation in long-term recovery. *Integrative Physiological and Behavioral Science* 31:289-305.

M Molnar, JE Skinner (1992) Low-dimensional chaos in event-related brain potentials. *International Journal of Neuroscience* 66:263-276.

M Molnar, G Gacs, G Ujvari, JE Skinner, G Karmos (1997) Dimensional complexity of the EEG in subcortical stroke – a case study. *International Journal of Psychophysiology* 25:193-199.

S Plymale (2004) The point correlation dimension: too good to be true? Term paper for *Introduction to Nonlinear Dynamics in Physiology*, The Johns Hopkins University.

J Pumprla, K Howorka, D Groves, M Chester, J Nolan (2002) Functional assessment of heart rate variability: physiological basis and practical applications. *International Journal of Cardiology* 84:1-14.

G Sammer (1999) Working memory load and EEG-dynamics as revealed by point correlation dimension analysis. *International Journal of Psychophysiology* 34:89-101.

JE Skinner, C Carpeggiani, CE Landisman, KW Fulton (1991) Correlation dimension of heartbeat intervals is reduced in conscious pigs by myocardial ischemia. *Circulation Research* 68:966-976.

JE Skinner, CM Pratt, T Vybiral (1993) A reduction in the correlation dimension of heartbeat intervals precedes imminent ventricular fibrillation in human subjects. *American Heart Journal* 125:731-743.

JE Skinner, M Molnar, C Tomberg (1994) The point correlation dimension: performance with nonstationary surrogate data and noise. *Integrative Physiological and Behavioral Science* 29:217-234.

JE Skinner, SG Wolf, JY Kresh, I Izrailtyan, JA Armour, MH Huang (1996) Application of chaos theory to a model biological system: evidence of self-organization in the intrinsic cardiac nervous system. *Integrative Physiological and Behavioral Science* 31:122-146.

JE Skinner, M Molnar (1999) Event-related dimensional reductions in the primary auditory cortex of the conscious cat are revealed by new techniques for enhancing the non-linear dimensional algorithms. *International Journal of Psychophysiology* 34:21-35.

JE Skinner, BA Nester, WC Dalsey (2000) Nonlinear dynamics of heart rate variability during experimental hemorrhage in ketamine-anesthetized rats. *American Journal of Physiology – Heart and Circulatory Physiology* 279:H1669-H1678.

PK Stein, MS Bosner, RE Kleiger, BM Conger (1994) Heart rate variability: a measure of cardiac autonomic tone. *American Heart Journal* 127:1376-1381.

RJ Storella, HW Wood, KM Mills, JK Kanters, MV Hojgaard, NH Holstein-Rathlou (1998) Approximate entropy and point correlation dimension of heart rate variability in healthy subjects. *Integrative Physiological and Behavioral Science* 33:315-320.

E Toth, I Kondakor, F Tury, A Gati, J Weisz, M Molnar (2004) Nonlinear and linear EEG complexity changes caused by gustatory stimuli in anorexia nervosa. *International Journal of Psychophysiology* 51:253-260.

H Tsuji, MG Larson, FJ Venditti Jr, ES Manders, JC Evans, CL Feldman, D Levy (1996) Impact of reduced heart rate variability on risk for cardiac events. The Framingham Heart Study. *Circulation* 94:2850-2855.

T Vybiral, DH Glaeser, AL Goldberger, DR Rigney, KR Hess, J Mietus, JE Skinner, M Francis, CM Pratt (1993) Conventional heart rate variability analysis of ambulatory electrocardiographic recordings fails to predict imminent ventricular fibrillation. *Journal of the American College of Cardiology* 22:557-565.

Chapter 20

Case study – Epidemiology

As far as nonlinear dynamical approaches are concerned, the field of epidemiology is slightly unusual, in that there is a relatively small number of studies almost all of which are of very high quality. We begin this chapter with a brief overview and background of dynamical approaches in this field, followed by a more detailed description of specific nonlinear forecasting approaches. Overviews of some of the major issues of dynamics in ecology, including spatial dynamics not discussed in this book, have been provided by May (1995) and Rai and Schaffer (2001). A review pertaining to population dynamics in general is also worthwhile (Bjørnstad, & Grenfell 2001). In this regard it is worth noting that chaotic dynamics have been demonstrated in a laboratory population of flour beetle (Cushing *et al.* 2001).

20.1 Background and early approaches

There is a long and distinguished history of mathematical modeling in this field (see May 1995 for short review). Among the first studies to look for deterministic explanations for seemingly random variations in childhood disease epidemics (measles, chickenpox, mumps) is that of Yorke and London (London & Yorke 1973, Yorke & London 1973). They estimated contact rates (fraction of susceptible individuals that an infective person exposes in a given time period) from monthly reporting of disease cases, and used these rates as one parameter of a simple model. Their results implied a possible deterministic cause for systematic variations in outbreaks in large populations, while random effects could explain these variations in smaller populations.

In a theme that we see repeatedly throughout works on deterministic chaos, the authors of one review point out the appeal of a deterministic dynamical model versus a completely stochastic one (Hamilton et al. 1997): "Therefore, deterministic models hold promise for greater realism in description, more parsimonious explanation and for providing a more authentic model with which to test hypothetical control techniques."

Solidifying the concept that a simple deterministic model could generate random-appearing variability in such data, May (1976) showed that the single-parameter logistic equation can have very complex dynamics. Recall from Chapter 1 that the logistic equation, or map, is often used as a simple model of population dynamics:

$$x(n+1) = \mu x(n)[1 - x(n)].$$

Here, $x(n)$ is the population in season n, and μ is a parameter that can take on a value between 1 and 4. "Population" here could also refer to a proportion of individuals in a population who show symptoms of a given disease. See section 1.4 for a discussion of the interpretation and some of the properties of this equation, which has been studied extremely widely (see Strogatz 1994, among many others). Depending on the parameter μ, the dynamics of $x(n)$ can result in a fixed point, periodicity, or even chaos. This was explained beautifully in May's classic work.

In the context of the state-space approaches that are of more interest to us, Schaffer (1985) was among the first to look at field data (Canadian lynx population) in terms of phase portraits. When the reconstructed attractor is sliced by a plane, the resulting trajectory intersections (the Poincaré map: see Chapter 12) lie approximately along a line. This implies that the dynamics on the section are one dimensional, and a first-return map that predicts the location of an intersection along this line, based on the previous intersection point, is single-humped and roughly triangular. Yearly peaks in the number of lynx trapped for fur show a similar first-return map. The equation $x(n+1)=ax(n)^2\exp(-bx(n))$, when fit to the map, indicates a change in dynamics of the fitted functions (a bifurcation), possibly due to a change in trapping efforts, in the early 1800s.

Similar analyses were then applied to time series of monthly measles cases in New York City (Schafer & Kot 1985a, 1985b), with roughly similar results. The attractor trajectory flowed mainly along a curved two-dimensional surface, and again the Poincaré section indicated one-dimensional dynamics with a single-humped first-return map, suggesting a dimension of two to three for the attractor overall. While we discuss a specific extension of these results below, other recent developments can be found in the edited volume by Perry and colleagues (2000).

The implications of these studies for epidemiological research are many, but one that demonstrates the change in viewpoint that is engendered by considering dynamical concepts is the following: "Field experiments ... remain valid only if one knows at the outset where the system happens to be in phase space. When one considers that the time scale for ecological attractors often exceeds that of research grants, the difficulties in interpretation become obvious" (Schaffer & Kot 1985a).

20.2 Nonlinear forecasting of disease epidemics

In order to address questions of periodicity (seasonal variation), randomness, and nonlinear dynamics, Sugihara and May (1990) carried out a series of studies using nonlinear forecasting to characterize the dynamics of various disease epidemics. In addition to devising a simplified forecasting technique, they were able to identify differences in dynamics between different diseases, and to relate these differences to known biological properties of the diseases.

The two main data sets in their first study consisted of monthly reports of measles infections in New York City from 1928 to 1963 ($N=432$), and monthly reports of chickenpox in the same location from 1949 to 1972 ($N=532$). All time-series data were first-differenced to remove linear trends, and hence to make the data more dense on the attractor (essentially identical results were obtained without this step).

Forecasting was carried out with the method that was outlined briefly in section 7.3 of Chapter 7. The time delay L for the attractor reconstruction was set to 1, and embedding dimensions M from at least 1 to 10 were used. For a given starting point, or reference point, the $M+1$ nearest neighbors were found which form the smallest simplex (i.e., the neighbors must surround the reference point, and have the smallest diameter of any such set). These neighbors were followed p steps ahead in time, and the location of the reference point within this new projected simplex was determined, based on exponential weighting of the distances from the neighbors to the reference ("weighted center of mass"). This forecasting was carried out for a range of reference points and forecasting steps. The correlation coefficient quantifies the correspondence between actual and forecast values, as a function of forecasting step p. Typically, each time series was separated into two halves, and forecasts were carried out on the second half based on the template of points in the first half (i.e., reference points were selected from the second half, nearest neighbors from the first half).

One secondary observation was that the one-step-ahead ($p=1$) forecasting quality often deteriorated as M increased. This is presumably because, as M increases and the number of nearest neighbors increases, not all of the nearest neighbors are necessarily close to the reference point, and they detract from the forecasting quality. This presents another way – possibly more robust than the correlation dimension – to estimate the approximate dimensionality of the reconstructed attractor. (See description of the related DVS technique in section 11.5.)

The authors noted that colored noise presents a problem in interpretation, and they suggest that comparison to a linear predictor, which should do worse for chaotic dynamics, would help distinguish noise from chaos. (See section 7.6 for a more promising development on this topic, which arose after the epidemiology work was published.)

For the measles data, the one-step forecasting quality was best at an embedding dimension M of 5 to 7, consistent with an attractor dimension of 2 to 3 (since embedding dimension must be at least $2D+1$ for a proper reconstruction, at least in theory: see Chapter 4). This is consistent with previous work that also suggested a dimension between 2 and 3 (Schafer & Kot 1985a, 1985b). Fig. 20.2.1 shows forecasting quality for these data, using data from the first half to forecast the second half (solid line), and also using the same data for both reference points and neighbors (dashed line). Similarity between the two sets of forecast suggests that there are no significant temporal trends (the data are stationary). The forecasting quality starts high and decays rapidly with time step, consistent with chaotic dynamics. In addition, forecasting with the method described above was better than that from a linear forecasting method (fitting a linear difference equation to the entire data set), again indicating the presence of nonlinearity.

For the chickenpox data, one-step forecasting quality was best with M equal to 5 or 6. Fig. 20.2.2 shows good, near-constant quality, forecasting across time steps, consistent with non-chaotic determinism such as a strong seasonal cycle. Here, the linear predictor does about as well as the method described above. These results together suggest that chickenpox exhibits noisy periodic dynamics. (Refer to Chapter 7 for a review of forecasting and its interpretation.)

These results were interpreted in light of the known properties of measles and chickenpox. Measles has a high reproductive rate, and an infected individual is infectious for a time but then immune and non-infectious for life. The resulting intrinsic dynamics, interacting with annual patterns (such as transient increases in population density during school semesters) can lead to two-year cycles and other complex dynamics (Grenfell *et al.* 1994). Chickenpox is less reproductive, and the major dynamics are due to seasonal periodicity (e.g., school semesters).

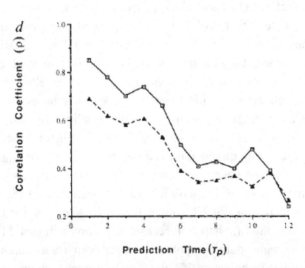

Figure 20.2.1. Nonlinear forecasting of monthly New York City measles data, suggesting chaotic dynamics. (Reprinted by permission from Macmillan Publishers Ltd: *Nature* 344:734-741, copyright 1990 (Sugihara & May, Fig. 4d).)

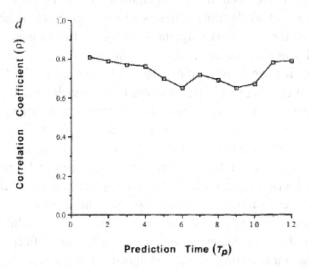

Figure 20.2.2. Nonlinear forecasting of monthly New York City chickenpox data, suggesting noisy seasonal dynamics. (Reprinted by permission from Macmillan Publishers Ltd: *Nature* 344:734-741, copyright 1990 (Sugihara & May, Fig. 5d).)

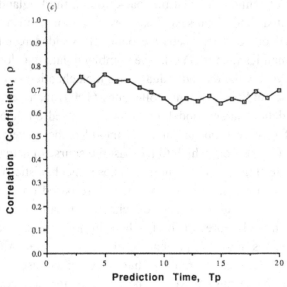

Figure 20.2.3. Nonlinear forecasting of England and Wales measles data, suggesting noisy seasonal chaotic dynamics. (From Sugihara *et al.* 1990, Fig. 5c, with permission.)

Extending this work, the same analysis was applied to measles time series data from England and Wales (Sugihara *et al.* 1990). These were smaller data sets and so the time series were not segmented as above. Unlike the results for New York City, the dynamics in this case appeared to reflect a noisy and deterministic but non-chaotic cycle (Fig. 20.2.3). Why then the difference between these data and those from NYC? One possibility is that the England/Wales data are from a much more broadly dispersed population, and we must now consider the spatial dynamics of the disease. To investigate this, the authors created a simple model consisting of a combination of different cities, each with disease dynamics represented by a logistic equation plus sinusoidal seasonal variation. Each logistic equation had a different parameter μ. When the time series from each city were added together, the chaotic (logistic) components were "averaged out" as uncorrelated noise, while the seasonal sinusoids were retained, yielding results that reflect noisy periodicity. The more terms (cities) included in the sum, the less effective was nonlinear forecasting relative to linear forecasting, reflecting a relative increase in the linear (periodic) dynamics versus the nonlinear chaotic dynamics. Against this background, the England/Wale data were separated into data sets from seven major British cities; forecasting carried out on these separate data sets yielded results that reflect chaotic dynamics (as for NYC). The combined data thus look like a noisy periodic cycle while the individual cities look chaotic.

Based on these studies, Tidd and colleagues (1993) made extensive use of surrogate data (random models) and models based on theoretical considerations of disease propagation. This included the clever use of model data as a template from which to forecast the course of actual data, as a means of testing a given model. These mechanistic models performed better than the stochastic ones that were based on surrogate data. Although convincing evidence of chaotic dynamics was not presented, these models forecast best when in the chaotic regime, suggesting that chaos may be present in the actual data. Although forecasting quality decayed so rapidly that its practical usefulness is questionable, the forecasting of yearly peaks is better and could be beneficial in planning vaccination schedules. Further work along these lines (Grenfell *et al.* 1994) addressed some of these issues, and examined the effects of embedding parameters and vaccinations.

Ellner and colleagues (Ellner *et al.* 1995) pointed out some problems with these types of dynamical studies in epidemiology, and made some suggestions for improvements. Specifically, they recommended the generation of appropriate surrogates. The choice of a proper surrogate for these data is not trivial, especially since these time series may have strong seasonal trends (periodicity). An improved surrogate is created by adding a seasonal trend to linearly filtered Gaussian white noise (possibly with a transformation). A number of measures were used in this study, including forecasting in reverse time. The overall conclusion was that measles does indeed have nonlinear dynamics, although other claims of chaotic disease dynamics could not be validated.

In confirmation of these studies, Olsen and Schaffer (1990) fit parameters to the standard SEIR (susceptible-exposed-infected-recovered) model for chickenpox and measles, and demonstrated from the model data that chickenpox is well-modeled as a periodic variation, while seasonal measles variations are chaotic. The correlation dimensions of the measles model data ranged from 2 to 3.

We close by noting that, given some understanding of the dynamics of a given population, chaos control techniques (Chapter 12) might be applicable to stabilize desired dynamics. This could be done through system perturbations by such methods as immigration control and (for animal populations) selective trapping and hunting (e.g., Constantino *et al.* 1997, Doebeli & Ruxton 1997, Gamarra *et al.* 2001, Hudson *et al.* 1998). The effects of random noise on the dynamics make the implementation of control methods such as these far from routine.

References for Chapter 20

ON Bjørnstad, BT Grenfell (2001) Noisy clockwork: time series analysis of population fluctuations in animals. *Science* 293:638-643.

RF Constantino, RA Desharnais, JM Cushing, B Dennis (1997) Chaotic dynamics in an insect population. *Science* 275:389-391.

JM Cushing, SM Henson, RA Desharnais, B Dennis, RF Costantino, A King (2001) A chaotic attractor in ecology: theory and experimental data. *Chaos, Solitons and Fractals* 12:219-234.

M Doebeli, GD Ruxton (1997) Controlling spatiotemporal chaos in metapopulations with long-range dispersal. *Bulletin of Mathematical Biology* 497-515.

S Ellner, R Gallant, J Theiler (1995) Detecting nonlinearity and chaos in epidemic data. In: D Mollison (ed) Epidemic Models: Their Structure and Relation to Data. Cambridge: Cambridge University Press.

JGP Gamarra, RV Solé, D Alonso (2001) Control, synchrony and the persistence of chaotic populations. *Chaos, Solitons and Fractals* 12:235-249.

BT Grenfell, A Kleczkowski, SP Ellner, BM Bolker (1994) Measles as a case study in nonlinear forecasting and chaos. *Philosophical Transactions of the Royal Society of London, Series A*, 348:515-530.

P Hamilton, JE Pollock, DAF Mitchell, AE Vicenzi, BJ West (1997) The application of nonlinear dynamics in nursing research. *Nonlinear Dynamics, Psychology, and Life Sciences* 1:237-261.

PJ Hudson, AP Dobson, D Newborn (1998) Prevention of population cycles by parasite removal. *Science* 282:2256-2258.

WP London, JA Yorke (1973) Recurrent outbreaks of measles, chickenpox and mumps. I. Seasonal variation in contact rates. *American Journal of Epidemiology* 98:453-468.

RM May (1976) Simple mathematical models with very complicated dynamics. *Nature* 261:459-467.

RM May (1995) Necessity and chance: deterministic chaos in ecology and evolution. *Bulletin of the American Mathematical Association* 32:291-308.

LF Olsen, WM Schaffer (1990) Chaos versus noisy periodicity: alternative hypotheses for childhood epidemics. *Science* 249:499-504.

JN Perry, RH Smith, IP Woiwod, DR Morse (eds) (2000) Chaos in Real Data: Analysis of Non-Linear Dynamics from Short Ecological Time Series. New York: Springer.

V Rai, WM Schaffer (2001) Chaos in ecology. *Chaos, Solitons and Fractals* 12:197-203.

WM Schaffer (1985) Order and chaos in ecological systems. *Ecology* 66:93-106.

WM Schaffer, M Kot (1985a) Do strange attractors govern ecological systems? *Bioscience* 35:342-350.

WM Schaffer, M Kot (1985b) Nearly one dimensional dynamics in an epidemic. *Journal of Theoretical Biology* 112:403-427.

S Strogatz (1994) Nonlinear Dynamics and Chaos. New York: Addison-Wesley.

G Sugihara, B Grenfell, RM May (1990) Distinguishing error from chaos in ecological time series. *Philosophical Transactions of the Royal Society of London. Series B, Biological Sciences* 330:235-251.

G Sugihara, RM May (1990) Nonlinear forecasting as a way of distinguishing chaos from measurement error in time series. *Nature* 344:734-741.

CW Tidd, LF Olsen, WM Schaffer (1993) The case for chaos in childhood epidemics. II. Predicting historical epidemics from mathematical models. *Proceedings: Biological Sciences* 254:257-273.

JA Yorke, WP London (1973) Recurrent outbreaks of measles, chickenpox and mumps. II. Systematic differences in contact rates and stochastic effects. *American Journal of Epidemiology* 98:469-482.

Chapter 21

Case study – Psychology

We end this section of the book, on case studies, with an area in which some of the applications of nonlinear dynamical approaches must be regarded as speculative. We begin with an overview of the appeal of nonlinear dynamics for psychology, and then discuss in more depth the dynamics of mood swings and schizophrenia symptoms, and the ability of humans to predict future values of chaotic sequences. Readers interested in pursuing this general area further might want to peruse some recent copies of the journal *Nonlinear Dynamics, Psychology, and Life Sciences*, published by the *Society for Chaos Theory in Psychology and Life Sciences*.

21.1 General concepts

The ideas of nonlinear dynamics and chaos have great appeal for those attempting to place psychological constructs on a mathematical foundation. The idea that seemingly complex behavior can be exhibited by a rather simple system, following deterministic and therefore explicable rules, provides hope that often convoluted and inexplicable human behavior might also follow some comprehensible rules. The dynamical notions of stability, complexity, and especially chaos, are appealing metaphors for the constant change and apparent self-organization that are frequently seen in the behavior of individuals and groups. These behaviors exhibit many of the qualities that we have seen in chaotic systems: changes in response that are not proportional to changes in a control variable, uncertainty and unpredictability, and sensitivity to initial conditions such that repetition of identical stimuli does not lead to identical responses. In clinical settings especially, there is a strong desire for testable quantitative models, in order to aid in the design of effective strategies for patient treatment.

Guastello (2001) has outlined in general form some of the ways in which nonlinear dynamical concepts might come into play in the various subfields of psychology. Among these ideas is the generation of creative solutions to a given problem through chaotic dynamics – what might otherwise appear to be randomly generated candidate solutions may instead be thought of as chaotic (and therefore unpredictable) outcomes of a deterministic system (human thought processes or group dynamics).

It is as yet unclear as to how reliable many of the studies in this area are, especially those in patients which are necessarily limited in data quality and quantity. To quote from a published abstract: "A review of how chaos theory is used in psychology reveals two relatively distinct efforts: chaos as a mathematical model of psychological phenomena and chaos as a metaphor for psychological phenomena. A discussion of recent articles reveals that most chaotic analysis fails to respect the minimum qualifications for data subjected to such analysis. Further, uses of chaos as an analogy for psychological phenomena are rife with misunderstandings of chaos" (Kincanon & Powel 1995).

With this sobering thought in mind, we now discuss two areas in which nonlinear dynamical approaches have been applied with at least some semblance of rigor, where the dynamical approach can provide possibly useful new interpretations of a psychological phenomenon, and where these interpretations might lead to testable hypotheses.

21.2 Psychiatric disorders

It is potentially important to understand the dynamics and causes of the time course of different psychological and psychiatric symptoms, so that appropriate treatments can be based on the underlying causes. Two prevalent models for mood swings in bipolar disorder suggest either an inherent periodicity, or a steadily increasing frequency as abnormal episodes become more spontaneously triggered with disease progression.

Gottschalk *et al.* (1995) designed a study to examine the dynamics of bipolar disorder, to see if chaos or other dynamical models would provide a better explanation for observed mood variations than would these two existing models. Time series in this study consisted of average daily mood reports, on a scale of 1 to 100, from seven patients and 28 control subjects. Qualitative observation of these self-report data suggests that there is neither a dominant periodicity nor a steadily increasing rate of occurrence of mood swings, although the patients do show intermittent episodes of periodicity. There are obvious differences between the patients and the normal control subjects, with more changes in pattern in the patients. Visual examination of two-dimensional state space (time-delay plots) also show a difference between normals and patients.

There are broadband frequency spectra from the data in both groups, although the spectra are flatter in normals, indicating long-term correlations, which might be either chaotic or random (see Chapter 13).

Correlation dimension estimates converged only for six of the seven patients; that is, the dimension estimate reached a plateau as embedding dimension increased. This was true for none of the normal control subjects. Patient dimensions ranged from 1.1 to 4.8. Although this is a troublingly large range if attempts to model the underlying dynamics are to be made, at least dimension estimates could be found for the patients. This is the main result of the study, the implications of which are discussed below.

A number of procedures were used to validate these findings. The authors checked for variation in the dimension as a function of the time delay (L) used in the attractor reconstruction. Although relatively constant, the dimension in most patients spanned an integer value, which the authors rightly interpreted to mean that the dynamics are not necessarily chaotic but might instead reflect noisy periodicities or quasi-periodicities (see Chapter 1). They also checked the data for stationarity, visually with recurrence plots, to ascertain that there was not significant change due to patient treatment over the course of the examined time series. Three types of surrogates were also tested (see Chapter 6): random-shuffle, phase-randomization, and amplitude-adjusted Fourier transform (AAFT). In all cases the surrogates did not yield dimension estimates: there was no convergence with increasing embedding dimension.

While the exact dimension values may be questionable due to the data acquisition methodology (daily self-reporting) and the small data sets (several hundred points), there nevertheless appears to be a clear difference between the patients and the control subjects in this study. In particular, there is a more organized temporal structure in the bipolar patients. The meaning of this is unclear, but it could reflect altered coupling between internal oscillators, or between internal processes (such as circadian rhythms) and external stressors. The possible existence of chaotic dynamics in this disorder raises the enticing prospect that nonlinear forecasting and chaos control strategies could be useful as assessment and treatment strategies.

This appealing result was later called into question by Krystal and Greenside (1998), who cited a theoretical study that suggests that a truly chaotic system should exhibit a region of exponential (rather than power-law) spectral decay. The original authors' response was that variability in the spectral estimates did not allow the acceptance or rejection of either spectral model.

Roughly similar results have been found in schizophrenia patients (Tschacher *et al.* 1997). In this study of 14 patients, the data consist of daily staff ratings of symptom levels on a 7-pont scale. Each time series is 200 to 770 points. Since the data are limited in amount and resolution, the authors purposefully avoided dimension estimation and instead used nonlinear forecasting methods (modified versions of the methods presented in Chapter 7). They also forecast randomized surrogate data, phase-shuffled surrogate data, and data from a linear autoregressive model fit to each time series. Of the 14 patients, eight had evidence of nonlinear dynamics, four of linear dynamics (ability to forecast data from the corresponding linear model), and two of randomness (near-constant poor forecasting). Nonlinearity, when present, was suggestive of chaotic dynamics: short-term forecasting ability that decayed rapidly. In supporting work, the authors showed that forecasting could indeed be carried out on such small discretized data sets. They did not, however, examine the decay rate of forecasting quality as a means to distinguish between chaos and randomness (section 7.6 in Chapter 7), nor demonstrate that such a distinction could be made with limited data of the type that they analyzed.

In apparent contrast with this result, which suggests a lower "complexity" (better forecasting), and hence possibly lower dimensions, in schizophrenics relative to normals, Koukkou *et al.* (1993) found increased dimensionality of the EEG in schizophrenics. This lower-level phenomenon (recording of neural activity) may reflect activation and separation of different neuronal assemblies, leading to difficulty in establishing a cohesive mental picture. This could then lead to a decreased dimension of higher-level processes which would reflect a consequent inability to respond coherently to stress or other external stimuli (Tschacher *et al.* 1997).

Other studies have noted impaired temporal processing in schizophrenia (see Paulus & Braff 2003), which might reflect either chaotic or random dynamics, but nevertheless indicates impaired temporal structure as a feature of the disease process.

21.3 Perception and action

Human behavior is rife with unpredictability. Even so, it is well known that humans cannot generate random sequences reliably (Wagenaar 1972). This apparent reluctance to deal with true randomness is also evidenced by the well-known "gambler's fallacy": humans expect that a long losing series "should be" shortly followed by a win in order to maintain randomness, while in fact for a truly random game of chance the past history has no effect on the outcome of subsequent trials (Ward & West 1998). This raises the question of whether or not apparently unpredictable human behavior might arise from chaotic dynamics in neural and psychological processes. This in turn leads to the question: are humans "sensitive to chaos"? Specifically, can they reproduce or forecast a chaotic sequence, better than an appropriate random control sequence?

A number of recent studies have examined the ability of humans to forecast or generate chaotic sequences. As we will see, even if subjects do not always match the desired chaotic process, they often produce a sequence that has nonlinear deterministic structure. A low-dimensional chaotic mechanism might be one way for the brain to generate unpredictable behavior, which can have such advantages as engendering creativity, aiding in problem-solving by generating non-obvious solutions, and avoiding enemies through evasive actions (Neuringer & Voss 1993).

In one of the first studies in this area (Neuringer & Voss 1993), subjects were asked to predict the future locations of a point along a line segment; the locations were governed by the chaotic logistic map (Chapter 1), so that although the sequence of positions might appear random, it was in fact completely deterministic. Error feedback (the difference between predicted and actual locations) was provided on each trial. Subject performance in this task was improved in a second set of trials, evincing a possible learning of the chaotic dynamics during the first set. Furthermore, the one-step-ahead predictions made by the subjects matched the general form of the logistic equation or map (see Fig. 12.2.1 for a depiction of this equation, which maps values from one time step to the next). The simplest interpretation of this result is that subjects could learn simple chaotic dynamics.

Metzger (1994) questioned this interpretation, suggesting instead that the results could be due to paired-associate learning, in which subjects learned approximate stimulus-response pairs, without any need to approximate or detect an underlying set of dynamics. Could the human prediction results, in other words, simply reflect a heuristic learning approach?

To address this question, in a subsequent study (Ward & West 1998), subjects were again asked to forecast position along a line, controlled by the one-step-ahead logistic map, with error feedback on each trial. After a set of learning trials, subjects were given a starting value and asked to iterate several steps ahead without feedback, in an attempt to reproduce the learned map. Delay-time plots show that subjects could produce maps that resembled the logistic function, but not exactly. Equations fit to the reproduced maps yielded values for the logistic equation parameter μ that correspond to a limit cycle (periodic behavior), rather than to the actual chaotic dynamics asked for in the learning sessions. A computational forecasting method due to Casdagli (1992), and described in section 11.5, was applied to the subjects' iterated predictions, and showed that their predictive behavior likely has a nonlinear deterministic component and is not completely random. Although a noisy logistic model reproduced some of the subject results, a fuzzy memory-pair model was even better. In this latter model, learned memory pairs (a given value and the subsequent predicted value) were modified by adding noise, in effect suggesting that subjects learned fuzzy groupings of sets of data rather than precise pairs. Thus, the learning of chaotic dynamics *per se* was likely not an explanation for the results, given the good performance of this alternative model. However, the presence of nonlinear determinism, as revealed by nonlinear forecasting, does suggest that there is a nonlinear deterministic process operating during the task, and that this process might be chaotic and could even be the source of the noise terms. Overall, the results suggest that the learning of a chaotic map can be accomplished at least in part with a heuristic approach.

More recently, Heath (2002) showed subjects eight values from a Hénon system, and had them predict the subsequent four values, with no error feedback. (These data were, however, heavily processed: rescaled and truncated.) Prediction ability was compared to that from an AAFT surrogate (Chapter 6), to see if human prediction ability is based on linear stochastic correlations in the data. Prediction of the chaotic data was better than that of the surrogate, indicating that the human prediction is "sensitive to chaos," although a heuristic learning pattern could not be ruled out.

A similar study by Smithson (1997) is interesting because of its implications for human decision making. Subjects were asked to forecast, one step ahead, both persistent and anti-persistent (see Chapter 13) nonlinear deterministic processes (chaotic), and random processes with the same distributions (created by shuffling the order of the values in each sequence). Prediction performance with the chaotic sequences was better, with greater accuracy and less under-dispersion. "Less under-dispersion is important because it indicates that subjects are less likely to under-estimate the extreme fluctuations in a chaotic process than they are in a random one, thereby rendering them better prepared for extreme outcomes.... these results suggest that our judgmental heuristics may have been shaped by nonlinear dynamical processes rather than stochastic ones, and evolved accordingly." In other words, the problem that humans have in generating random sequences, and in erroneously believing the gambler's fallacy, may result from inexperience with truly random processes, which have independent trials. Rather, heuristic rules may have developed through exposure to natural processes that do exhibit correlations, either persistent or anti-persistent, and possibly deterministic and chaotic.

Finally, Gilden *et al.* (1995) demonstrated that the error sequence in estimating spatial or temporal intervals has a $1/f$ form. This does not necessarily indicate the presence of chaos but is related to concepts in Chapter 13 on temporal structure and long-term correlations.

In closing, recall from Chapter 4 that an attractor can be reconstructed from discrete-event spike-train data, which will reflect the underlying continuous dynamics that trigger the spikes (with some reasonable assumptions on the integrate-to-fire mechanism). Thus it might be better to examine patient and other psychological data, which has been highly discretized (into a small number of categories), as discrete events, the times of which coincide with the rating exceeding a certain critical value.

References for Chapter 21

M Casdagli (1992) Chaos and deterministic versus stochastic non-linear modeling. *Journal of the Royal Statistical Society B* 54:303-328.
DL Gilden, T Thornton, MW Mallon (1995) $1/f$ noise in human cognition. *Science* 267:1837-1839.
A Gottschalk, MS Bauer, PC Whybrow (1995) Evidence of chaotic mood variation in bipolar disorder. *Archives of General Psychiatry* 52:947-959.
SJ Guastello (2001) Nonlinear dynamics in psychology. *Discrete Dynamics in Nature and Society* 6:11-29.
RA Heath (2000). Nonlinear Dynamics: Techniques and Applications in Psychology. Mahwah, NJ: Erlbaum.
RA Heath (2002) Can people predict chaotic sequences? *Nonlinear Dynamics, Psychology, and Life Sciences* 6:37-54.
E Kincanon, W Powel (1995) Chaotic analysis in psychology and psychoanalysis. *Journal of Psychology* 129:495-505.

M Koukkou, D Lehmann, J Wackermann, I Dvorak, B Henggeler (1993) Dimensional complexity of EEG brain mechanisms in untreated schizophrenia. *Biological Psychiatry* 33:397-407.

AD Krystal, HS Greenside (1998) Low-dimensional chaos in bipolar disorder? *Archives of General Psychiatry* 55:275-276.

MA Metzger (1994) Have subjects been shown to generate chaotic numbers? Commentary on Neuringer and Voss. *Psychological Science* 5:111-114.

A Neuringer, C Voss (1993) Approximating chaotic behavior. *Psychological Science* 4:113-119.

MP Paulus, DL Braff (2003) Chaos and schizophrenia: does the method fit the madness? *Biological Psychiatry* 53:3-11.

M Smithson (1997) Judgment under chaos. *Organizational Behavior and Human Decision Processes* 69:59-66.

W Tschacher, C Scheier, Y Hashimoto (1997) Dynamical analysis of schizophrenia courses. *Biological Psychiatry* 41:428-437.

W Wagenaar (1972) Generation of random sequences by human subjects: a critical survey of the literature. *Psychological Bulletin* 77:65-72.

LM Ward, RL West (1998) Modeling human chaotic behavior: nonlinear forecasting analysis of logistic iteration. *Nonlinear Dynamics, Psychology, and Life Sciences* 2:261-282.

Chapter 22

Final remarks

Readers who have made it to this point, having read through the case studies, may have a growing sense of frustration or disappointment, stemming from the fact that it is hard to find definitive results in the area of nonlinear dynamics applied to physiological systems and data. Investigations are often equivocal, with more or less indication of nonlinearity or determinism, which seems to change with the next generation of studies. This reflects the relative youth of the field and the difficulty in studying living systems.

Nevertheless some progress has been made. Perhaps one of the most important points of progress is a change in viewpoint – the recognition that complex dynamics and apparent randomness can result from simple underlying rules, and that these rules and the resulting dynamics might change as a result of pathology. Closely related to this is the notion of "healthy variability" and "dynamical diseases," the latter of which might reflect a decrease in "complexity" or degrees of freedom (dimensionality).

A second very important consequence of the recent emphasis on nonlinear dynamics is the development of computational tools, which enable the investigation of hypotheses based on these new viewpoints. This book has emphasized these computational tools, although hopefully the last several chapters have given some sense of how the tools might be applied in the context of these new views of dynamics and physiology.

Many studies in this field look like the proverbial "fishing expedition," wherein a large number of computational tools are applied to a system of interest, to see if any interesting results are obtained. While typically the death knell for a grant proposal, this is an approach that must be taken seriously in this field, given the possible artifacts that can arise from the use of any single procedure.

There is some irony in the fact that some of the most successful applications of nonlinear dynamics to physiology have made use of the concepts behind the computational tools, as opposed to the strict definitions of the underlying mathematics. Examples include the identification of eye-movement fast phases by fitting to a high-dimensional plane (Chapter 4), and the use of the cross-correlation integral to detect the onset of epileptic seizures (Chapter 18). The latter, in particular, is a good example of a case in which the precise meaning of the dynamical measure is unclear; it is not a dimension *per se*, but a measure of "dynamical similarity," which can encompass quite a lot of effects.

There are of course many other fields of study which have begun to feel the probing of nonlinear dynamics approaches. These are beyond the scope of the present book, but brief list will give an idea of the broad applicability of the concepts and the range of fields in which traditional linear methods are reaching their limits. These fields include economics and finance, seismology, fluid turbulence, and meteorology. In the case of weather, an interesting series of articles appeared in the journal *Nature* wherein a scholarly debate of sorts ensued over the validity and interpretation of low-dimensional attractors derived from climate data. (Citations to these articles appear below.)

One of the chief goals of this book is to provide the reader – whether student, teacher, or researcher – with sufficient knowledge to be able to read and critically assess the growing literature on nonlinear dynamics. This is especially important in life sciences applications, which are saddled with the problem of having limited amounts of noisy data. Even if the reader does not apply these computational tools in his or her own work, it can still be of great value to be able to judge the validity of the claims made in the research literature. It is of course also the author's wish that the material in this book would encourage readers to try these methods in their own research.

References on climatic attractors

AA Tsonis, JB Elsner (1988) The weather attractor over very short timescales. *Nature* 333:545-547.
C Essex, T Lookman, MAH Nerenberg (1987) The climate attractor over short timescales. *Nature* 326:64-66.
P Grassberger (1986) Do climatic attractors exist? *Nature* 323:609-612.
EN Lorenz (1991) Dimension of weather and climate attractors. *Nature* 353:241-244.
C Nicolis, G Nicolis (1984) Is there a climatic attractor? *Nature* 311:529-532.

Suggested references for further study

There are several important topics that could not be covered in this book, with its emphasis on didactic presentation of the most standardized techniques. Some of these topics are spatiotemporal chaos, Lyapunov exponents, and recent improvements in the design of surrogate data. In the future, one can expect to see further refinement of the computational techniques covered here (and others), with a continuing emphasis on applications to short, noisy, nonstationary data sets. Two areas that appear especially ready for new developments are the use of nonlinear dynamical measures as a way to test and refine mathematical models, and the objective incorporation into the computational tools of pre-existing information on the system under study.

For those interested in pursuing further some of these ideas, as well as those covered in detail in this book, the following books are recommended.

HDI Abarbanel (1996) Analysis of Observed Chaotic Data. New York: Springer-Verlag. Overview of theory and techniques, with somewhat the same flavor as the present volume.
RH Abraham, CD Shaw (1992) Dynamics: The Geometry of Behavior. New York: Addison-Wesley. Beautiful picture-book of dynamics and attractors, with mathematical commentary.

PS Addison (1997) Fractals and Chaos: An Illustrated Course. Philadelphia: Institute of Physics Publishing. Excellent general introduction to the mathematical aspects of nonlinear systems.

JB Bassingthwaighte, LS Liebovitch, BJ West (1994) Fractal Physiology. Bethesda MD: American Physiological Society. Describes the application of fractal concepts in physiology.

DJ Bell (1993) Mathematics of Linear and Nonlinear Systems. New York: Oxford University Press. More mathematical than the other recommended texts but still accessible. Places dynamical systems into the context of set theory, vector spaces, and other mathematical structures.

GA Edgar (1990) Measure, Topology, and Fractal Geometry. New York: Springer-Verlag. Provides rigorous mathematical background for understanding fractal geometry.

K Falconer (2003) Fractal Geometry: Mathematical Foundations and Applications, 2nd edition. Hoboken NJ: John Wiley & Sons. Provides rigorous mathematical background for understanding fractal geometry, with applications.

GW Flake (1998) The Computational Beauty of Nature: Computer Explorations of Fractals, Chaos, Complex Systems, and Adaptation. Cambridge MA: MIT Press. A clever book covering the topics listed in its subtitle. Fun and educational. See the web site at http://mitpress.mit.edu/books/FLAOH/cbnhtml/home.html

D Kaplan, L Glass (1995) Understanding Nonlinear Dynamics. New York: Springer. Undergraduate-level text that introduces basic aspects of dynamical thinking.

BB Mandelbrot (1983) The Fractal Geometry of Nature. New York: WH Freeman and Co. A true classic by the "inventor" of fractals.

E Ott (2002) Chaos in Dynamical Systems, 2nd edition. New York: Cambridge University Press. Another fine mathematically oriented text.

E Ott, T Sauer, JA Yorke (1994) Coping with Chaos: Analysis of Chaotic Data and The Exploitation of Chaotic Systems. New York: Wiley-Interscience. Volume of reprints of early papers in the field.

S Strogatz (1994) Nonlinear Dynamics and Chaos. New York: Addison-Wesley. Excellent mix of mathematical theory, geometrical intuition, and application examples. A perfect starting point for those with a basic background in college mathematics.

JR Weeks (2002) The Shape of Space, 2nd edition. New York: Marcel Dekker. A beginning topology book with a strong emphasis on the intuition behind the mathematics.

Appendix

MATLAB Example Code

This appendix presents a few simple examples of how some of the computational tools presented in this book can be implemented in computer code. We have chosen to use the MATLAB™ software platform, since it is easy to learn and use, and ubiquitous in signal-processing applications. Even if you are unfamiliar with MATLAB, the close correspondence between the code and the underlying mathematics should make these examples instructive.

The examples here are presented mainly for initial explorations and to develop intuition. More advanced applications would require some modification and extension of these routines. The reader is of course welcome to modify and augment these samples as desired. As an alternative, at least one commercial software package is available for performing some of these computational tasks, and an internet search will quickly yield a number of sites where dynamicists are willing to share their own software for specific computations.

Please note that these program examples come with no guarantees as to their correctness or suitability for any particular use. They are intended as educational tools only. Most of these programs are rudimentary, in that they do not check for invalid or nonsensical input values, nor will they provide error messages. (Note that % indicates that the following text is a non-executable comment, in the MATLAB code.)

A.1 State-space reconstruction

dplot simply generates a two-dimensional state-space plot, using $x(i)$ and $x(i+d)$.

```
function dplot(x,N,d,tp)
% dplot(x,N,d,'type')
% Plot array x on x-axis, delayed version of x on y-axis.
% Delay is by d points.
% 'type' is line type: '-', 'o', etc.
plot(x(1:N),x(1+d:N+d),tp)
grid
return
```

dplot3 is identical to **dplot** except that it is in three dimensions. The MATLAB **view** command can be used to change the viewing angle.

```
function dplot3(x,N,d,tp)
% dplot(x,N,d,'type')
% Plot array x on x-axis, delayed version of x on y-axis,
%   double-delayed version of x on z-axis.
% Delay is d points.
% 'type' is line type: '-', 'o', etc.
plot3(x(1:N),x(1+d:N+d),x(1+2*d:N+2*d),tp)
grid
return
```

dplot3d is similar to **dplot**, except that two versions of the state-space attractor are plotted, one in red and one in green. One version is rotated (by the argument *theta*) about the vertical axis, creating a stereogram. A three-dimensional view can be produced by viewing the graph with red/green spectacles (a red lens over one eye and a green lens over the other). An offset of *theta*=3 should work well.

```
function dplot3d(xx,N,d,theta)
% dplot3d(xx,N,d,theta)
% Plot array x on x-axis, delayed version of x on y-axis.
% Delay is d points.
% Also plot rotated version,
%   rotated by theta degrees around vertical axis.
x=xx(1:N);
y=xx(1+d:N+d);
z=xx(1+2*d:N+2*d);
```

```
x2=x + z.*tan(theta*pi/180);
plot(x,y,'r')
hold
plot(x2,y,'g')
hold off
return
```

dplot3d2 is similar to **dplot3d**, except that the two versions of the state-space attractor are plotted next to each other. A three-dimensional view can be produced by placing an index card between the two graphs, resting your nose on the card, and defocusing as if looking at a distance. The two graphs should merge into a single view with a perception of depth. (A stereoscope may also be used.)

```
function dplot3d2(xx,N,d,theta)
% dplot3d2(xx,N,d,theta)
% Plot array x on x-axis, delayed version of x on y-axis.
% Delay is d points.
% Also plot adjacent rotated version,
%   rotated by theta degrees around vertical axis.
x=xx(1:N);
y=xx(1+d:N+d);
z=xx(1+2*d:N+2*d);
x2=x + z.*tan(theta*pi/180);
subplot(231)
plot(x,y)
axis square
subplot(232)
plot(x2,y)
axis square
return
```

makea1 is the basis of several other routines. It generates the "trajectory matrix" **A** (see Chapter 4). This matrix contains the points of the reconstructed attractor along its rows.

```
function A=makea1(x,N,M,J,L)
% Create trajectory matrix.
```

```
% A=makea1(x,N,M,J,L)
% x=data vector
% N=number of points on trajectory (rows)
% M=embedding dim: number of putative state variables (columns)
% J=increment in starting values for each attractor point (typically 1)
% L=time delay (lag)
for i=1:M
 A(:,i)=x(1+(i-1)*L:J:1+(N-1)*J+(i-1)*L);
end;
return
```

FFN1 computes the proportion of False Nearest Neighbors in the attractor represented by matrix **A** (from **makea1**), over the range of embedding dimensions from 2 to *M* (where *M* is the embedding dimension used in the creation of the matrix **A**). A distance ratio (when going from dimension *M* to *M*+1) greater than *Rtol* defines a false neighbor; a value of about 10 is a good place to start.

```
function pffn=FFN1(A,Rtol)
% pffn=proprtion of false nearest neighbors
% A=trajectory matrix (from makea1)
% Rtol=tolerance level for detecting neighbors
%
% initialize values
N=size(A,1);
Mmax=size(A,2);
pffn=zeros(1,Mmax-1);
%
% cycle through the embedding dimensions
for M=1:Mmax-1
  % pick reference point
  for iref=1:N
  ref=A(iref,:);
  % go through all other points and compare to reference
  %   to find nearest neighbor
  dst=inf;
  for j=1:N
```

```
    dd=0;
     for k=1:M
     dd=dd+(A(iref,k)-A(j,k))^2;
     end
    dd=sqrt(dd);
     if(j~=iref & dd<dst) dst=dd; inear=j; end
    end
    % for this M and iref, nearest neighbor index is inear, distance is dst
    % calculate distance at next higher M
    dst2=sqrt(dst^2 + (A(iref,M+1)-A(inear,M+1))^2);
    if(dst2/dst > Rtol) pffn(M)=pffn(M)+1; end
    end % end iref loop
   end % end M loop
   pffn=pffn/(N-1);
   return
```

A.2 Correlation dimension

This is perhaps the most dangerous of the samples in this appendix, since as noted in Chapter 5 reliable computation of the correlation dimension involves a great deal of care and attention. Nevertheless, this routine will allow for the exploration of some simple systems. It produces correlation integrals; the slope of the correlation integral is an estimate of the dimension, and this step is left to the reader, as are the myriad other details outlined in Chapter 5. This correlation dimension routine is slow, with an execution time proportional to N^2; faster algorithms are available, and execution can also be faster if the algorithm is written in a language such as C. Note that the program uses the *maximum norm* to define distance, because of its speed advantage (to find the distance between two vectors using the maximum norm, subtract corresponding elements of the vectors, and take the magnitude of the largest of the resulting values). To change the values of embedding dimension M, delay time L, and number of points N, specify these parameters when making the matrix **A** with **makea1**. The distances over which the correlation integral is to be computed are in the array r; a logarithmically spaced set of values can easily be produced in MATLAB (for example, this command will produce 100 values ranging from 0.1 to 100, that is, from 10^{-1} to 10^2: **r=logspace(-1,2,100)**.) The values of the correlation integral are returned in array *Cr*, and can be plotted with: **loglog(r,Cr)**. The "Theiler window" *Twin* was defined in Chapter 4, and is the time period over which consecutive points are not considered in the computations, in order to avoid counting pairs of points with spurious temporal correlations in the integral.

```
function Cr=DcorrBasic(r,skip,A,Twin)
% Calculate number of points on trajectory within
% certain distance of given reference point, and repeat
% using evenly-spaced reference points, separated by skip
% (i.e. to use all points as reference, set skip=1).
% Cr=correlation integral estimate array
% r=distance criterion array
% skip=increment in choosing reference points
% A=trajectory matrix (each row is a point on the trajectory)
% Twin="Theiler window" length
%
[N,M]=size(A);
rsize=length(r);
k=zeros(rsize,1);
nn=N/skip;              % number of ref points
iref=0;
kount=0;
%
% Loop to pick reference point
for j=1:skip:N
AA=A-ones(N,1)*A(j,:);    % subtract point j from others
AA=max(abs(AA')); % distance from j to others (max norm)
AA(max(1,j-Twin):min(N,j+Twin))=...
   inf*ones(1,length(max(1,j-Twin):min(N,j+Twin)));
% substitute inf for pts in Twin
      kount1=N-length(max(1,j-Twin):min(N,j+Twin));
kount=kount+kount1;
iref=iref+1;
```

```
%
    AA=sort(AA');
    AA=AA(find(AA~=inf));
%
    i1=find(r<min(AA));
    i2=find(r>max(AA));
    k(i2)=k(i2)+kount1*ones(size(i2))';
%
    i2=min(i2);
    i=max(i1); if(isempty(i)) i=1; end
    for ii=1:length(AA)
            if(AA(ii)>r(i))
                    k(i)=k(i)+ii-1;
                    i=i+1;
                    if(i==i2|i>length(r)) break; end;
            end
    end
end;
Cr=(k')/kount;
return
```

A.3 Surrogate data

The simplest surrogate consists merely of shuffling the data points, which can be done conveniently with this routine.

```
function y=shuffle(x)
% Random shuffle of the input values.
i=rand(size(x));
[i,is]=sort(i);
y=x(is);
return
```

A more sophisticated surrogate is created by randomizing the phases of the frequency components in the frequency spectrum, which is accomplished by this routine. Randomization of the time series can be verified by plotting the input and output: **y=phase_rand1(x); plot(x); plot(y)**. The fact that the power spectrum has been retained can be verified by plotting the two power spectra: **Fx=fft(x); Fy=fft(y); subplot(211); loglog(abs(Fx(1:end/2)).^2); subplot(212); loglog(abs(Fy(1:end/2)).^2)**.

```
function xx=phase_rand1(x)
[n1,n2]=size(x);
if(n2~=1) x=x'; end
N=length(x);
fx=fft(x);
fx=fx(1:round(1+N/2));              % positive freq only
ph=2*pi*rand(round(1+N/2),1);       % get new phases
ph=exp(ph*sqrt(-1));                % convert to complex
ph(1)=1;                            % don't change DC component
fxx=abs(fx).*ph;                    % apply new phases to old magnitudes
fxx=[fxx(1:round(1+N/2))' conj(fxx(round(N/2):-1:2))']';
% reassemble with odd-symmetry phase
ifxx=ifft(fxx);
xx=real(ifxx);
return
```

A.4 Forecasting

NLF performs nonlinear forecasting on the attractor in trajectory matrix **A**. See the comments in the program code for a description of the input arguments and the method. For a given reference point (starting point), the nearest neighbors are selected from the first *Ntemp* points. There are *irefn* reference points, starting at point *iref1* (*iref1* must be greater than *Ntemp*). The program uses a subroutine that appears below, and the final results are determined and plotted with the final routine in this section.

```
function [pred,err]=NLF(A,Ntemp,iref1,irefn,k,nstep)
% Find prediction error for 1 to nstep steps ahead.
% Method: for each subsequent prediction with the
%   same reference point, treat the previous prediction
```

```
%    as an actual point, and repeat the prediction
%    procedure (iterative prediction).
% pred(i,j)=prediction from ref pt i, at j steps ahead
% err(i,j)=prediction error (i=ref point, j=steps ahead)
% A=trajectory matrix (each row is a point)
% Ntemp=nr of template points from which to choose predictors
% iref1=first reference point (predictee), >Ntemp
% irefn=number of reference points
% k=number of neighbors to use in each prediction
% nstep=maximum steps ahead to predict at each predictee point
%
if(iref1<=Ntemp)
disp('Error -- iref1 must be > Ntemp');
return;
end
%
[N,M]=size(A);
if(N<iref1+irefn+nstep)
disp('Error -- not enough points in A matrix')
return
end
%
for i=1:irefn
iref=i+iref1-1;
AA=A;
% do this loop for each ref pt
for j=1:nstep
            Aref=AA(iref,:);                    % ref pt
            nextref=iref+j-1;
            if( j>1) AA(nextref,:)=Aref; end;
            [idx,dst]=NearNbr(AA,iref,k,1,Ntemp);
% find k nearest neighbors
            grp1=AA(idx,:);                     % domain
    idx2=idx+j;                         % project j steps ahead
    grp2=AA(idx2,:);                    % range
    coeff=grp1\grp2;                    % L-S solution
```

```matlab
        guess=Aref*coeff;
        pred(i,j)=guess(M);
                Aref=guess;            % base next pred on this pred
        actual=AA(iref+j,:);
        err(i,j)=norm(guess-actual);   % distance error
    end;
end;
return
```

This subroutine finds nearest neighbors for the forecasting routine above.

```matlab
function [idx,dst]=NearNbr(A,jref,k,ifirst,ilast)
% Calculate the k nearest neighbors to the point on the
% trajectory that is indexed by jref.
% Neighbors must be between indices ifirst and ilast.
% A=trajectory matrix (each row is a point on the trajectory)
% idx=indices of the k nearest neighbors
% dst=distances
%
% Set reference point
xref=A(jref,:);
% Truncate trajectory; only include pts between ifirst and ilast
NN=ifirst:ilast;
AA=A(NN,:);
% Compare points to reference point
AA=AA-ones(max(size(AA)),1)*xref;
AA=AA.^2;
AA=sum(AA')';
% Let's search dist^2, otherwise put this in: AA=AA.^(.5);
[dst,idx]=sort(AA);
dst=dst(1:k).^(.5);      % return distances, not squared
idx=idx(1:k);
return
```

Having computed the predicted values *pred(i,j)* from reference point with index *i*, to *j* steps ahead, now find the correlation coefficient when comparing the predicted and actual values at each forecasting step *j*. This routine returns the value of this correlation coefficient for $j=1$ to *nstep* steps ahead. It assumes that the value J was equal to 1 when the matrix **A** was generated by **makea1**. To see the results: **plot(rho)**.

```
function rho=CorrPred(x,pred,iref1,M,L);
[irefn nstep]=size(pred);
ML=(M-1)*L;
for k=1:nstep
   aa=corrcoef(x(iref1+ML+k:iref1+ML+irefn+k-1), pred(:,k));
   rho(k)=aa(1,2);
end
return
```

The forecasting routines can be tested with a deterministic sine wave as follows:

```
s=sin(1:1000);
As=makea1(s,900,3,1,3);
[preds,errs]=NLF(As,300,301,300,5,10);
rhos=CorrPred(s,preds,301,3,3);
plot(rhos)
```

Forecasting of a random signal can be tested as well:

```
r=randn(1000,1);
Ar=makea1(r,900,3,1,3);
[predr,errr]=NLF(Ar,300,301,300,5,10);
rhor=CorrPred(r,predr,301,3,3);
plot(rhor)
```

A.5 Recurrence plots

RecMat generates a recurrence matrix (matrix of inter-point distances) from a trajectory matrix **A** (see **makea1** above).

```
function RM=RecMat(A)
[N,M]=size(A);
for j=1:N
   AA=A-ones(N,1)*A(j,:);
   AA=sqrt(sum(AA'.^2));
```

```
    RM(j,1:N)=AA;
  end
  return
```

Having produced a recurrence matrix with **RecMat**, the next routine will normalize the distances and apply a threshold (0-1), then create the resulting recurrence plot.

```
function RecPlt(RM,thr)
% RM=recurrence matrix
% thr=threshold (0-1)
% Distances less than thr are dots, greater than thr are white.
% Distance is scaled from 0 to 1 and thr is based on that scale.
a=RM;
amin = min(min(a));
amax = max(max(a));
a = ((a-amin)/(amax-amin));
idx=zeros(size(a));
idx(find(a>thr))=0;
idx(find(a<=thr))=1;
cmap=[1 1 1; 0 0 0];
colormap(cmap);
imagesc(idx);
set(gca,'ydir','normal');
axis square
return
```

A.6 Periodic orbits

The identification of periodic points involves, at its simplest, finding attractor points that are close together (recurrent points). This routine does that, taking as its input the trajectory matrix **A**. The argument *gap* tells the program not to look at points closer together in time than this value, so that points along a single trajectory path do not appear as recurrent simply because they are close in time.

```
function Porbits(A,tol,N,gap)
```

```
% A=trajectory matrix
% tol=distance tolerance that defines recurrence
% N=number of points to examine
% gap=don't test points closer together than this
%
for iref=1:N-gap
for itest=iref+gap:N
    Aref=A(iref,:);
Atest=A(itest,:);
if(norm(Aref-Atest)<tol) m(iref)=itest-iref; break; end
end
end
hist(m,100)
return
```

A.7 Poincaré sections

poinb generates a Poincaré section from the attractor in trajectory matrix **A**. The coordinates of the points on the section are given by (a,b). It is assumed that the attractor is three-dimensional so that the section is two-dimensional.

```
function [a,b]=poinb(A,N,plane,val,dir)
% Poincare section.
% Plot points that cross plane determined by 'plane'=val.
% (i.e. plane='x' and val=10 plots all points y and z where
% x=10), going in direction specified by 'dir'.
% Interpolates to get plane-crossings.
% A=trajectory matrix
% dir: positive=through plane from - to + (positive-going)
% dir: negative=through plane from + to - (negative-going)
%
% Get values from the trajectory matrix.
x=A(1:N,1); y=A(1:N,2); z=A(1:N,3);
% Find indices where trajectory crosses plane.
pl=eval(plane);
```

```
pl=pl-val; pl=sign(pl); pl=diff(pl);
%
if(dir>0) ijk=find(pl==2); end
if(dir<0) ijk=find(pl==-2); end
% pl=abs(pl);
%
if(plane=='x')
apre=y(ijk); apst=y(1+ijk);
bpre=z(ijk); bpst=z(1+ijk);
cpre=x(ijk); cpst=x(1+ijk);
end
%
if(plane=='y')
apre=x(ijk); apst=x(1+ijk);
bpre=z(ijk); bpst=z(1+ijk);
cpre=y(ijk); cpst=y(1+ijk);
end
%
if(plane=='z')
apre=x(ijk); apst=x(1+ijk);
bpre=y(ijk); bpst=y(1+ijk);
cpre=z(ijk); cpst=z(1+ijk);
end
%
del=(val-cpre)./(cpst-cpre);
a=apre+del.*(apst-apre);
b=bpre+del.*(bpst-bpre);
plot(a,b,'o')
return
```

A.8 Software packages

MATLAB
The MathWorks, Inc.
3 Apple Hill Drive
Natick, MA 01760-2098
508-647-7000
General mathematics and numerical analysis program, very extensive, a *de facto* standard for signal processing. Highly recommended.
www.mathworks.com

Mathcad
Mathsoft Engineering & Education, Inc.
101 Main Street
Cambridge, MA 02142
Fax: 617-444-8007
800-628-4223
Another general mathematics package in wide use.
www.mathsoft.com

Chaos analysis software
Developed by Mike Banbrook
Department of Electrical Engineering
University of Edinburgh
Free software, designed for a graphical interface (Xwindows), written in C. Includes: Time series embedding, Mutual information, Singular value decomposition, Lyapunov exponents, Poincaré sections.
http://www.see.ed.ac.uk/~mb/analysis_progs.html

Chaos Data Analyzer
Physics Academic Software
940 Main Campus Drive
Suite 210
Raleigh, NC 27606-5212.
800-955-8275
http://webassign.net/pas/index.html
Student and professional versions available. The same organization has a number of other chaos and dynamics related programs for education and research. Includes: Probability distributions, Polynomial fitting, Power spectra, Hurst exponent, Lyapunov exponent, Capacity dimension, Correlation dimension, Correlation function, Correlation matrix, Phase-space plots, Return maps, Poincaré movies, Nonlinear prediction, Surrogate data.

Dataplore
ixellence GmbH
+49 (0) 33 75 - 508 616
+49 (0) 7 21 - 151 553 084
info@ixellence.com
Commercial package. Includes: Delay plots, Statistics, Surrogate data, Fourier analysis, Linear filtering, Wavelets, Recurrence plots, Correlation dimension, Lyapunov exponents.
http://www.ixellence.com/

Recurrence Quantification Analysis (RQA)
CL Webber
Department of Physiology
Loyola University Chicago
2160 South First Avenue
Maywood, IL 60153 USA
http://homepages.luc.edu/~cwebber/

Time Series Analysis: TISEAN
R Hegger, H Kantz, T Schreiber
Institut für Physikalische und Theoretische Chemie, Universität
Frankfurt (Main), and Max-Planck-Institut für Physik komplexer
Systeme, Dresden
Popular free package for time-series analysis, extremely extensive.
http://www.mpipks-dresden.mpg.de/~tisean/TISEAN_2.1/index.html
R Hegger, H Kantz, T Schreiber (1999) Practical implementation of
nonlinear time series methods: The TISEAN package. *Chaos* 9:413-435.

TSTOOL
C Merkwirth, U Parlitz, I Wedekind, W Lauterborn
Free MATLAB-based package.
Includes: Time-delay reconstruction, Lyapunov exponents, Fractal
dimension, Mutual information, Surrogate data, Poincaré sections,
Nonlinear prediction and forecasting.
http://www.physik3.gwdg.de/tstool/

Visual Recurrence Analysis (VRA)
E Kononov
Free package with many routines, derived mostly from recurrence
analysis. Includes: Recurrence plots, Recurrence quantification analysis,
Average mutual information, False nearest neighbors, Correlation
dimension, Nonparametric modeling.
http://pweb.netcom.com/~eugenek/download.html

Keeping in mind the caveats noted at the beginning of this appendix, the
author might be persuaded to share electronic versions of his routines if
contacted by email (mjs@dizzy.med.jhu.edu, as of date of publication).

A.9 Sources of sample data sets

Those readers who wish to try their hand at the analysis methods covered in this book – or who wish to develop their own – are fortunate that there are many internet archives of data on which to test these techniques. This is a great resource for the research community and the author salutes those investigators who maintain these sites and make their data so freely available.

A quick search of the World Wide Web for "chaotic data sets" will provide many such archival sources. Some are maintained specifically for access by researchers, while others are meant to provide data for university-level homework assignments. In addition, many published papers will list an internet archive of data used in the paper.

A particularly good site for physiological data (and many other computational resources) is PhysioBank, at the PhysioNet web site: http://www.physionet.org/.

Random numbers provide an important test case for many algorithms. Probably the most reliable random numbers that are easily available are those generated from radioactive decay times, available at: http://www.fourmilab.ch/hotbits/.

Index

1/f spectrum, 217, 240, 309

AAFT (see amplitude adjusted Fourier transform surrogate)
acetylcholine, 276
adaptability, 277
aging, 16, 276
amplitude adjusted Fourier transform surrogate (AAFT), 110, 112, 115, 133, 138, 230, 283, 304, 308
amplitude-reversal, 133, 134
angst, 78
anoxia, 288
anticontrol, 208, 209
antifibrillatory, 277
anti-persistent, 212, 243, 308
apamin, 137, 138
aperiodic, 175
arrhythmias, 277, 286
artifacts, 286
astronomy, 14
Atanasoff, John, 3
attractor, 4, 15, 16, 46, 48, 54-60, 65-70, 73-75, 80-96, 104, 113, 141-150, 155, 157, 159, 160, 166, 181, 182, 184, 185, 190, 194, 196, 200, 202, 203, 211, 215, 226, 227, 230-232, 237, 246-248, 264, 265, 268, 269, 293, 294, 295, 304, 309, 312, 317-319, 328

attractor reconstruction, 100, 151, 254, 256, 278, 318
aura, 263
autocorrelation function, 30, 34, 38-43, 56, 107, 109, 112, 116, 135, 212, 213, 216, 229, 230, 256
autonomic, 276, 277, 288
autoregressive, 133, 220, 253, 282, 305
autoregressive moving average (ARMA), 133, 220
average, 20-24, 29
axonal propagation, 252

basal ganglia, 253
basis functions, 199, 221
beat-to-beat variation, 276-289
Bell Labs, 13, 14
Belousov-Zhabotinski reaction, 64, 176
beta-blocker medication, 280, 281
bifurcation, 194, 293
big bang, 14
binomial distribution, 170
bipolar disorder, 303-305
bispectrum, 254
blood pressure, 276
bootstrap, 104
botany, 211
box-counting dimension, 82
brain, 203, 217
brain function, 285

brainstem, 253
Brown, Robert, 211
Brownian motion, 15, 211, 212
butterfly effect, 4

Canadian lynx, 293
Cantor set, 77, 79, 80
cardiac, 152, 276-291
cardiac dysrhythmia, 240
cardiac rhythms, 179
cardiovascular disease, 276-289
Cartesian coordinate system, 63
center frequency, 152
center-of-gravity, 242
center-of-pressure (COP), 242
cerebellum, 253
cerebral hemispheres, 172
chaos, 2-4, 9, 22, 13-17, 43, 80, 94, 99, 101, 104, 115, 124, 125, 134-138, 148, 150, 187, 190, 191, 203, 208, 239, 240, 253, 254, 257, 293, 295, 298, 302, 303, 305, 306, 308, 309
chaos control, 178, 179, 190, 199, 202-208, 265, 299, 305
chaotic, 156, 164, 171, 175, 187, 190, 191, 194, 196, 203, 211, 214, 215, 246, 252, 256, 257, 260, 263, 278, 283, 288, 292, 293, 299, 302-308
chaotic attractor, 203, 209
chaotic data sets, 333
chaotic dynamics, 252, 292, 295, 298, 306
chaotic sequence, 306-309
chaotic system, 203
chickenpox, 292, 294, 296, 299
childhood disease, 292
circadian rhythms, 305
climate, 312
clinical diagnosis, 260
closed-loop control, 243
coarse-grained, 183, 184
coast of Britain, 75

coastlines, 76
cognition, 179
cognitive state, 262
coherent oscillatory state, 179
collinear, 208
colored noise, 29, 134, 295
communication, 156
complex dynamics, 293, 296
complexity, 73, 95, 149, 264, 266, 272, 276, 281, 288, 302, 306, 311
computer code, 316-333
confidence intervals (for dimension estimates), 92-94, 104
continuous, 51, 52, 61, 68, 155, 157, 169, 170, 181, 182
continuous-time, 171, 193, 196
contravariant basis vectors, 205
convulsions, 262
coronary artery, 286
correlated, 236, 239, 243
correlated noise, 138
correlation coefficient, 24, 25, 28, 30, 33, 34, 128, 132, 134, 135, 155, 281, 283, 295, 325
correlation dimension, 81-101, 106, 108, 114, 118, 121, 124, 126, 214, 215, 229, 230, 234, 239, 240, 246, 248, 253, 254, 257, 260, 263-265, 269, 278-280, 283-285, 295, 304, 320, 331, 332
correlation integral, 84-87, 90, 92, 97, 100, 229, 231, 254, 257, 265, 268, 269, 281, 283-285, 289, 320
correlation time, 56, 89-91, 93
cortex, 171, 253
cortical, 262, 271, 273
coupled systems, 156, 162, 171
coupling, 151, 155-172, 231, 237, 288, 305
covariance, 24
creativity, 307
critical, 217

cross-correlation, 30, 33, 34, 37, 116
cross-correlation integral, 269, 312
cross-recurrence plot, 165
cross-spectrum, 116

data sets, 333
degrees of freedom, 246-248, 263, 266
delay, 126, 136
delta-epsilon, 185
derivative, 194, 199
Descartes, 63
determinism, 141, 144, 150, 152, 160, 162, 169, 181, 185, 187, 296, 311
deterministic, 1, 8, 10, 11, 13, 14, 47, 56, 58, 63, 65, 110, 124-126, 138, 143, 148-151, 164, 169, 211, 213-216, 221-223, 226, 227, 230, 235, 239, 240, 243, 244, 246, 248, 278, 282, 292, 293, 298, 302, 303, 306-308
deterministic dynamics, 178, 243, 244, 245, 248
deterministic system, 143, 148, 149, 151, 175, 182, 183, 188
deterministic versus stochastic (DVS), 187, 188
detrended fluctuation analysis (DFA), 216
diagnostic, 276
diffeomorphism, 52
difference correlation coefficient, 136
difference equation, 6, 220, 296
differentiability, 157, 170, 181
differential equation, 2, 4, 6, 220
differential geometry, 50
diffusion coefficient, 242
dimension, 47, 48, 51, 54-60, 62, 65, 73-108, 114, 115, 117, 118, 121, 158-160, 170, 222, 242, 246, 247, 263-266, 268, 269, 312
dimensionality, 246, 248, 295, 311
disappointment, 311

discrete events, 309
discrete-time map, 194, 204
discrete-time, 171
disease, 240, 292-299, 303, 306
divergence of trajectories, 144, 148, 187
DNA, 152
dominant period, 215
dopamine, 252
dot product, 205, 206
drift, 114, 115, 143, 148-150, 234
dynamic system, 5, 7, 8
dynamical disease, 311
dynamical nonlinearity, 134
dynamical similarity, 268-271, 312
dyslexia, 240

economics, 312
edge of chaos, 217
EEG, 15, 16, 67, 95, 98, 115, 117, 171, 172, 178, 217, 218, 262-275, 306
effective dimension, 265-268, 271
eigenvalues, 200, 201, 205
eigenvectors, 193, 199-201, 205
Einstein, 211
EKG, 67, 115, 277-280, 287, 288
electrical stimulation, 263
embedding, 48-65, 73, 85-89, 96, 97, 142-144, 151, 157-160, 165, 167, 172, 214, 215
embedding dimension, 55, 56, 59, 65, 93-95, 98, 101, 128, 130, 132, 138, 142, 143, 151, 157-160, 172, 214, 230, 246, 247, 294, 295, 319
embedding window, 89, 91, 229
EMG, 121, 152
energy preservation, 288
Engineer's Prayer, 44
England, 297, 298
ENIAC, 3
ensemble, 21, 29
entrainment, 209

entropy, 149, 150
epidemics, 114, 137, 292-301
epidemiology, 292-301
epilepsy, 98, 171, 178, 179, 209, 240, 262-275, 312
error bars (for dimension estimates), 92-94, 104
error sequence, 309
essential tremor, 252-261
Euclidean dimension, 74, 75, 77, 78, 96
Euclidean distance, 182, 280
Euclidean space, 50, 51, 54
evasion, 307
Exceptional Events, 182, 185-187, 257
exercise, 276
existence (mathematical), 214
expected value, 21, 22
exponential, 135, 138, 214
eye-movement (also see oculomotor), 312

False Nearest Neighbors (FNN), 56-60, 68, 132, 151, 157, 158, 172, 247, 248, 319
feedback, 253
fibrillation, 179, 276-279, 283, 285-287, 289
filtered noise, 43, 94, 99, 134
filtering, 69, 99, 121, 234
finance, 64, 312
first-delay map, 196
first-differences, 114, 115, 280, 294
first-return map, 257, 293, 294, 279
first-step forecasting, 281, 282
fixed point, 11, 12, 190, 193-205, 207-209, 293
flow (trajectory), 198, 201
fluid turbulence, 312
forecasting, 124-140, 155, 156, 160-165, 172, 214, 215, 235, 279, 281, 282, 283, 305, 306-309

forecasting horizon, 137
Fourier, 33-38, 44
Fourier analysis, 33, 35, 38, 44
Fourier series, 38
Fourier spectrum, 213
Fourier transform, 35, 37, 55, 109, 111, 112, 116
fractal, 65, 75, 76, 135, 196, 197, 211, 212, 214, 218, 314
fractal attractor, 215
fractal dimension, 77
fractal time series, 211
fractional Brownian motion (fBm), 15, 34, 135, 211-216
fractional Gaussian noise (FGN), 213, 216
fractional integration, 212
frequency spectrum, 265, 304
frustration, 311
functions, 181
fuzzy memory-pair model, 308

gambler's fallacy, 306, 309
Gauss, 25
Gaussian, 40, 43, 44, 105-107, 109-112, 116, 184, 212, 214, 283
Gaussian noise, 264
Gaussian surrogate, 118, 121, 230
Gaussian white noise (GWN), 29, 40, 109, 111, 143, 212, 221, 253, 299
generalized synchrony, 156, 161, 162, 164, 165
genetic algorithm, 222
global fit, 132, 133
goodness of fit, 28
group dynamics, 303

hand, 252, 253, 256
harmonic oscillator, 45-48, 51-53, 95
healthy, 279, 280, 288
healthy variability, 16, 208, 311
heart attack, 277, 280, 282

heart rate, 276-289
heart transplantation, 279, 288
heart-beat intervals, 69, 208, 216
heart-rate variability (HRV), 276-291
Hénon attractor, 200
Hénon system, 162, 169, 171, 194, 197, 199, 201, 204, 205, 207, 285
heuristic learning, 307-309
hippocampus, 178, 209, 264
homeomorphism, 51-54, 75, 170
horizontal and vertical line segments (recurrence plot), 148
horizontal line segments (recurrence plot), 150
H-reflex, 121
human behavior, 302-309
hunting, 299
Hurst exponent, 212, 216
hypersynchronization, 262, 263

ictal, 262, 271
immigration, 299
increments, 213, 214, 216
independent trials, 309
inferior olivary nucleus, 253
information dimension, 83
information theory, 83, 149
injectivity, 170
input-output reconstruction, 68
instability, 177
instantaneous dimension, 279, 285
integral equation, 221
integrate-to-fire, 136, 309
inter-beat intervals, 98, 276-289
interdependence, 155-172
interictal, 271
internet, 214, 316, 333
inter-spike intervals, 117, 136
interventions, 263
invariant, 54
invertible, 156, 157, 161

ischemia, 286, 288
isometric transformation, 142

Jacobian, 158, 199, 200, 204
Jansky, Karl, 13
joint recurrence plot, 166-169
joint recurrences, 169

Kaplan, Daniel 96
ketamine, 288
Koch curve, 79-81

laminarity, 150
Laplace, 8, 43
learning of chaotic dynamics, 306-309
least-squares, 25, 131
Legendre, 25
limb, 242, 246, 252, 253
limb movement, 16
limit cycle, 12, 133, 209, 246, 247, 307
linear, 125, 127, 130-135, 138, 253, 254, 256, 257, 277, 282, 284
linear approximation, 193, 194, 200
linear correlation, 109, 111, 114, 116, 118, 155, 230, 268, 308
linear estimation, 133
linear forecasting, 296, 298
linear function, 24-25, 221
linear measures, 272
linear model, 118, 305
linear predictor, 283, 295, 296
linear process, 109
linear regression, 25-28, 127, 130, 131, 216, 220
linear system, 32-35, 38, 43, 44, 105, 107, 110, 132, 156, 193, 220
linear trend, 294
linearity, 8, 253
linearize, 193, 200, 201
linearly correlated, 111, 116, 118, 121, 134, 230, 264, 272

linearly filtered, 253, 299
Lissajous figure, 64
literature, 312
local approximation, 126, 131, 133, 283
local dimension, 247-248, 284
local dynamics, 247-248
local False Nearest Neighbors, 247-248
local interactions, 217
local linear approximation, 69, 125, 127, 132, 158
locomotion, 264
logarithm, 221
logarithmic compression, 110
logistic equation, 11, 125, 193, 197, 293, 298, 307
long-term correlation, 216, 227, 243, 304, 309
Lorenz attractor, 126, 183, 190
Lorenz system, 43, 58-60, 66-68, 86, 91, 118, 120, 133, 143, 165, 171, 184, 188, 215, 257, 260, 285
Lyapunov exponent, 65, 150, 186, 187, 313

Mandelbrot, Benoit, 75, 76
manifold, 50, 51, 54, 74, 201, 208
map, mapping (functions), 51, 53, 155, 169-171, 181, 182, 193-201, 204, 207, 307
mathematical model, 73, 220, 303
mathematical modeling, 226, 292, 313
mathematics, 1, 3, 17
MATLAB, 316-333
maximum norm, 320
mean, 20-24, 26, 27
measles, 117, 292-296, 298, 299
measure profile surrogate, 271
medication, 263
mental picture, 306
mental task, 264

metaphor, 303
meteorology, 312
model, 1-3, 6, 17, 221-223
mood, 302-306
mortality, 277
motoneuron, 253
motor control, 242-251
moving average, 121, 133, 220
multi-variable, 166, 167
multivariate recurrence plot, 167, 169
multivariate surrogate data, 115, 170
mumps, 292
muscle fatigue, 152
mutation, 222
mutual false nearest neighbors (MFNN), 157-160
mutual forecasting, 155, 160-165, 171, 237
mutual information, 56
mutual recurrence, 167

natural processes, 309
nearest neighbors, 127, 128, 130, 131, 157-162, 172, 182, 187, 247, 294, 295, 319, 323, 325, 332
neighborhoods, 157, 159, 161
nervous system, 217
neural dynamics, 179, 272
neural firing, 137, 262
neural firing patterns, 246
neural networks, 217
neural pathway, 253, 288
neural processes, 306
neural spike trains, 68, 69, 117, 136, 152, 178
neurocardiac, 288
neuromuscular, 242, 246
neuronal assemblies, 272, 306
neuroscience, 262
New York City, 293, 294, 298

NMDA receptors, 288
noise, 9, 11, 13-15, 50, 55, 56, 58, 65, 69, 70, 80, 87, 91, 92, 94, 101, 125, 133-138, 143, 151, 165, 278, 280, 281, 283, 295, 298, 299
noise radius, 114
noise reduction, 69, 70
nonlinear, 155, 160, 169-172, 211, 214, 252-254, 256, 257, 260, 283, 296
nonlinear dynamics, 254 276, 282, 294, 299, 302, 305, 311, 312
nonlinear forecasting, 64, 69, 124-140, 187, 223, 235, 280, 282, 283, 292, 294, 298, 305, 308, 323
nonlinear function, 221
nonlinear system, 109, 221
nonlinearity, 95, 101, 104, 109, 110, 117, 132, 133, 311
nonparametric, 221
nonrecursive, 234
nonstationarity, 213, 216, 267, 268, 279
nonstationary, 99, 126, 141, 143, 149-151, 279, 283, 313
noradrenaline, 276

observability, 50
observation function, 110
oculomotor, 39, 61, 68, 117, 121, 217, 223, 225-241
OGY (method of chaos control), 203, 204, 206-208
one-step forecasting quality, 132, 295, 296
one-step-ahead forecasting, 127, 130, 132, 135, 187
one-step-ahead predictions (human), 307
one-to-one, 170
open-loop control, 243
optimal observer, 156
optokinetic nystagmus (OKN), 39, 61, 117, 121, 223, 225-239

orthogonal, 205, 206, 208
oscillations, 252, 256, 257
oscillator, 253, 256, 257, 260, 305
out-of-sample, 128
output preservation, 288
oversampling, 254

paired-associate learning, 307
parallel flows, 183-185
parameter, 5, 7, 220-223
parametric, 221
parasympathetic, 276, 277
Parkinson's disease, 252
Parkinsonian tremor, 252-261
pathological, 252, 254, 256, 257, 280
pathology, 179, 208, 209, 216, 266, 276, 311
pathology (and dimension), 98
PD2 (see point correlation dimension)
Penzias, Arno, 14
percent determinism, 149, 152, 243
percent recurrence, 149, 150, 152
period, 143, 257
periodic, 132, 133, 138, 144, 148, 149, 151, 152, 191, 202, 203, 209, 225, 227, 234, 246, 293, 296, 298, 299, 307, 327
periodic components, 112
periodic orbits, 113, 175-179, 327
periodic surrogate, 231
periodicity, 12, 141, 176, 227, 294, 296, 298, 299, 303, 304
persistent, 212, 243, 308
perturbation, 178, 203, 205, 207, 208, 217, 299
phase spectrum, 109, 110, 116
phase-randomization surrogate, 109, 110, 112, 116, 118, 121, 230, 235, 237, 264, 278, 283, 304, 305, 332
phase-synchronization, 272, 273
physiologic tremor, 252-253
physiological, 223, 252, 253

physiological interpretation, 239
physiological state, 152, 265, 268
physiological system, 211, 311
physiology, 2, 5, 7, 14, 311, 312
piecewise linear, 130
Poincaré section (or map), 64, 65, 190-192, 199, 200, 202-205, 207, 257, 268, 278, 293, 294, 328
Poincaré, Henri, 2, 64
point correlation dimension (point D2, PD2), 283-289
point process, 136
point-wise dimension, 97, 284
pollen, 211
polynomials, 221
population dynamics, 292, 293
posture, 242-245, 276
potassium channels, 137
power law, 135, 213-217
power spectrum, 109, 111, 112, 116, 213, 230, 240, 253, 257, 323
power-law noise, 136
power-law scaling, 75, 76, 78, 81, 82, 86, 87, 90, 91, 97, 215, 284
power-law spectrum, 135, 213-215, 217, 305
predict, 276, 277, 302, 308
prediction, 124-141, 155, 160, 307
prediction, seizure, 265
predictive eye movements, 217
pre-seizure state, 271, 273
probability, 21-24, 29
problem-solving, 307
proportional, 302
protein structure, 152
pseudo-period, 229
pseudo-periodic, 113, 118, 120
pseudo-periodic surrogate, 118, 120
psychiatric disorders, 302-306
psychiatric, 303
psychology, 302-209

QRS complex, 277
quasi-periodicities, 12, 304

R wave, 277, 288
R-R intervals, 277
radio, 14
radioactive, 10
radioactive decay, 333
random, 47, 56, 65, 68, 105, 107, 109, 110-112, 114, 116-118, 125, 126, 135-138, 143, 144, 148, 169, 170, 172, 182, 183, 185-189, 211, 212, 214, 215, 221-223, 225, 230, 236, 239, 242-244, 248, 253, 256, 257, 260, 292, 293, 298, 299, 304, 306-308
random dynamics, 306
random fractal, 211, 215
random fractal sequence, 135
random noise, 188
random number generator, 9
random numbers, 333
random process, 21, 29, 30, 94, 185, 277
random sequences, 306, 309
random signal, 326
random system, 211, 214, 215, 221
random variables, 24, 29, 30, 105
random walk, 212, 242-243
randomization, 106, 107, 110, 112
randomness, 8, 9, 11, 13, 17, 95, 102, 104, 117, 217, 254, 256, 260, 278, 280, 283, 294, 305, 306, 311
range, 150, 151
Rapp, Paul, 95
Ratio (recurrence analysis), 150
reading, 239, 240
reconstruction, 158, 159
recruitment, 266, 268
recurrence analysis, 141-154
recurrence matrix, 142, 227, 326, 327

recurrence plot, 141-154, 165, 166, 169, 227, 243, 244, 304, 326, 327, 331, 332
recurrence plot matrix, 167
recurrence, 165-169, 176, 179
recurrence quantification analysis (RQA), 141, 149-152, 243, 244
recurrent, 142-144, 148-151, 327
recursive, 234
reinnervation, 288
Renyi dimensions, 83
reorganization, 280, 288
rescaled range, 216
research grants, 294
resonance, 247
respiration, 152, 276, 278
respiratory sinus arrhythmia, 278
rest, 264, 268
reverse time, 299
reversibility, 133, 134
rhythmic movements, 246-250
Robinson, David, 1
Romeo and Juliet, 7
roulette, 64

saccades, 217
saddle, 201
sample rate, 100, 229, 230
sampled data, 20-23, 28
sampling theorem, 100, 137
scale-invariance, 217, 218
scaling exponent, 243
scaling, 10, 32, 33, 214-218
scaling range (region), 86, 87, 91, 92, 100, 242, 264, 265, 281, 284
schizophrenia, 302, 305-306
scholar, 2
school semesters, 296
seasonal cycle, 296
seasonal trend, 299
seasonal variation, 294, 298

SEIR model (susceptible-exposed-infected-recovered), 299
seismology, 312
seizure, 262-273
seizure-time surrogate, 271
self-organization, 302
self-organized criticality (SOC), 217
self-organizing, 217
self-similar, 76, 79, 211, 212, 214, 217
sensitive dependence on initial conditions, 3, 13, 144, 187, 191, 302
sensorimotor, 217
set theory, 314
shock, 288
shuffle surrogate, 107, 108, 118, 121, 230, 304, 322
shuffle, 231, 264, 308
signal processing, 60, 61, 165, 316
similarity index, 268-271
simple harmonic motion, 12, 115
simplex, 131, 137, 172, 294
singular value decomposition (SVD), 100, 269
sleep, 263, 268, 286
smooth, 157, 161, 181
sojourn points, 151
solutions, 303, 307
spatial correlations, 141
spatial dynamics, 292, 298
spatial intervals, 309
spatial-temporal dynamics, 263
spatiotemporal chaos, 313
spatiotemporal correlations, 217
spectrum, 36-40, 43-44
spike trains, 137, 309
spike-density function (SDF), 137
spurious correlations, 254
stability, 176, 193, 199, 202, 217, 302
stabilize, 299
stable, 193-195, 200-203, 205, 208

stable manifold, 203, 205, 206, 208
stable orbit, 177
standard deviation, 23-24, 212, 216, 277, 280, 282, 286
state space, 4, 6, 45-71, 222, 293, 316, 317, 318
state variables, 5-7, 9, 45, 73, 95, 98, 101, 220, 235, 246
state vector, 63
state-space reconstruction, 45-72, 316
state-transition matrix, 63
static monotonic nonlinearity, 230, 283
static nonlinearity, 110-112, 115, 134
stationarity, 7, 34, 212, 227, 268, 277, 295, 304
statistical hypothesis testing, 105, 106
statistically independent, 24, 25, 29
stereo, 226
stereogram, 317
stereoscope, 318
stereoscopic, 226
stimulus-response, 307
strange attractor, 65, 73, 80, 94, 104, 175
Strangelove, Dr., 3
stress, 276, 286
stressors, 288, 305
stride intervals, 216
substantia nigra, 137
superposition, 10, 32
surface of section, 190
surgery, 263
surrogate data, 92, 93, 95, 97, 101, 104-123, 124, 132, 133, 138, 151, 170, 172, 179, 185, 215, 230, 235, 264, 271, 272, 278, 279, 283, 288, 298, 299, 304, 305, 313, 322, 331, 332
sway, 242-245
symmetry, 323
sympathetic, 276, 277, 278
synaptic transmission, 252

synchronization, 156, 157, 159, 263, 266, 272
synchronous, 262
system identification, 220-223

tachyarrhythmias, 277
Takens embedding theorem, 54, 66, 126, 285
tardive dyskinesia, 242
temporal changes, 279, 283
temporal correlations, 114, 135, 150, 229, 231, 237, 264, 272, 321, 327
temporal intervals, 309
temporal processing, 306
temporal trends, 295
test statistic, 106
thalamus, 253
Theiler correction, 90, 114, 115, 321
therapeutic intervention, 265
thought, 303
threshold distance (recurrence analysis), 142
time delay, 89, 128, 252
time-delay embedding, 45-72, 126, 128, 222
time-delay reconstruction, 45-72, 126, 128, 141, 157, 165, 240
time-interval data, 268, 269
time-reversal, 133, 134
timing jitter, 70
topological dimension, 74, 77, 78, 96
topological invariant, 126
topological property, 53, 54, 126
topological transformation, 155
topologically equivalent, 54, 55
topology, 3, 50, 142, 315
trajectory matrix, 130, 318, 328
transfer function, 35, 43
transient, 148, 149, 151
trapping, 293, 299
treatment, 303-305

tremor, 252-261
Trend (recurrence analysis), 150
t-test, 104, 105

Ulam, Stanislaw, 5, 19
uncertainty, 150, 302
uncorrelated, 25, 221, 232, 236, 272
uncorrelated noise, 113, 120, 143, 253
under-dispersion, 308
undersampling, 230
unequal time delays, 67, 68
unpredictable, 302, 303, 306, 307
unstable, 193, 194, 200, 201, 205, 206, 208
unstable periodic orbits (UPO), 175-180, 203, 327

vaccination, 298
vagus nerve, 278
van der Pol, 257, 260
variability, 208
variables, 4-7, 246
variance, 21-24, 26, 212

vector, 198, 200, 201, 205, 206, 208
vector fields, 183-185
vector spaces, 314
ventricular fibrillation, 97
vertical line segments (in recurrence plot), 150
vestibular, 225
Volterra-Wiener, 221
von Neumann, John, 3

Wales, 297, 298
walking, 216
weather prediction, 2, 4, 13, 312, 313
white noise, 29, 30
Whitney embedding theorem, 50, 54
Wigner, Eugene, 1
Wigner-Ville spectrum, 213
Wilson, Robert, 14
within-sample, 128
Wold decomposition theorem, 221
word occurrences in English, 152

zero-crossings, 268, 271